EARLY BRITISH COMPUTERS

EARLY BRITISH COMPUTERS

THE STORY OF VINTAGE COMPUTERS AND THE PEOPLE WHO BUILT THEM

Simon Lavington

Manchester University Press

Published by Manchester University Press, Oxford Road, Manchester
M13 9PL

British Library cataloguing in publication data

Lavington, Simon Hugh
Early British computers
1. Computers – History
2. Electronic data processing – Great Britain – History
I. Title
621.3819'5'0941 QA76

ISBN 0–7190–0803–4 hard covers
ISBN 0–7190–0810–7 paperback

Printed in Great Britain by
Unwin Brothers Limited
The Gresham Press, Old Woking, Surrey

CONTENTS

		page
1.	Introduction	1
2.	Computing in the 1930s	4
3.	The second world war	8
4.	The technology of early computers	13
5.	The ACE, the 'British national computer'	23
6.	The Cambridge EDSAC	31
7.	The Manchester Mark I	36
8.	The NPL Pilot ACE	44
9.	Transistor computers	48
10.	Defence computers	53
11.	Elliott Brothers	57
12.	Pioneering small computers	62
13.	Leo and English Electric	68
14.	Ferranti Ltd, ICT and ICL	78
15.	Programming an early computer	87
16.	Meanwhile in America . . .	98
17.	The National Research Development Corporation	102
Appendix 1:	the plain man's guide to computer terminology	106
Appendix 2:	technical specification of 15 early British computers	114
Appendix 3:	technical specification of 9 early American computers	120
Bibliography and references		126
Acknowledgements and picture credits		131
Index		133

1
INTRODUCTION

The word 'computer' prior to 1940 meant only one thing: a clerk equipped with a hand calculating machine who would 'compute' the standard calculations required for wages, actuarial tables, astronomical predictions, etc. This was of necessity a tedious process. From 1935 to 1945 the application of electronics to 'automatic computers' generally took the form of developing faster calculating equipment related to specific problems. However, in 1945 there arose an interest in the design of *universal* automatic computers which could be applied to an unlimited range of problems, once the appropriate 'program' had been fed into the machine. These universal computers are the ones we would recognise today.

The universal computer has become known as the *stored-program digital computer*, because it embodies two interesting characteristics not present in other types of calculating machine.

These characteristics are:

(*a*) An internal store (or memory) whose contents can be selectively altered automatically during computation. This store is used to hold both instructions (the program) and data.

(*b*) The ability of the machine to vary its actions in a strictly defined manner, according to the value of data items encountered during computation. In practice this is only achieved if data items are represented as numerical, i.e. digital, quantities inside the computer.

(For those unfamiliar with computer terminology, an introduction to the concepts and definitions of the more important technical terms is given in Appendix 1.)

Fig. 1.1 Part of Charles Babbage's Analytical Engine, designed between 1833 and 1837. This remarkable mechanical calculating machine was planned to have a store (or memory) for 1000 numbers, and was to be controlled by punched cards similar to those used in the Jacquard loom. In developing his engine, Babbage conceived the idea of what we would now call a *stored-program* computer.

The first person to appreciate the possibilities of stored-program computers was the Cambridge mathematician Charles Babbage (1791–1871), who planned a mechanical Analytical Engine which was to be controlled by punched cards.[1, 2] Babbage's implementation was, in the event, defeated by technical difficulties, but he introduced concepts which were remembered a century later once the electronic era had dawned.

Babbage apart, the question 'who invented the computer?' is really unanswerable. By the mid-1940s there were several research groups in Britain, America and Germany who were familiar with the basic concept of a stored-program digital computer. They worked from a basis of common sense, generally coupled with an intimate knowledge of existing special-purpose calculators such as the huge ENIAC (Electronic Numerical Integrator and Computer) – an American machine which is described later. The transition from calculators without the stored-program capability to those with it was a gradual though conscious one, related principally to the development of suitable storage techniques. The first machine based entirely on the stored-program principle was a very small computer which worked in Manchester University on 21 June 1948 but, as with the Wright brothers and the history of aviation, Manchester should be regarded as just one of several centres of activity.

This book centres on the emergence of an indigenous computer industry during the decade 1945–55. It is largely a story of individual effort seen against a background of various, usually unsuccessful, attempts at coordination on a national level. The decade in question saw an impressive crop of British ideas in spite of, or more likely because of, the lack of central coordination and lavish funding. All this, however, is to anticipate events. At the start of the period under discussion there were few people who knew (or cared) what a stored-program computer was. Even those who did were uncertain as to whether contemporary technology could be made reliable enough to give useful computing runs. To set the scene we must first consider what 'computing' meant before the second world war.

2
COMPUTING IN THE 1930s

Before the second world war there were three types of calculating machine in common use. Most numerous were the mechanical and electro-mechanical hand calculators which would add, subtract, multiply and divide two numbers. These machines were generally about the size of a typewriter. Most of them utilised mechanisms based on German, Swedish and American inventions. Secondly, there were electro-mechanical punched-card machines such as sorters and tabulators, derived from the pioneering work of Herman Hollerith and James Powers in America at the turn of the century. Originally applied to commerce and statistics,

Fig. 2.1 A popular type of hand calculator known as a Brunsviga, in recognition of its German origins. These machines could add, subtract, multiply and divide, and could be used to 'compute' tables of standard astronomical data, etc.

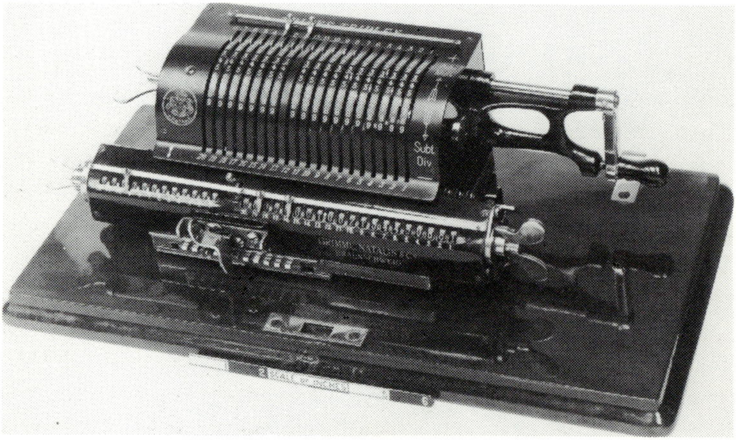

Fig. 2.2 Punched card equipment of the 1930s. Tabulators and sorters of the type shown above were widely used in commerce and administration. Note the Queen Anne legs!

punched-card devices began to be used for scientific calculations in the late 1920s. The third and most spectacular class of calculator was the Differential Analyser. This was a large and highly specialised 'equation solver'.[3]

The fundamental mechanism of a Differential Analyser is a wheel-and-disc integrator – see Fig. 2.4. This was invented in 1876 by James Thomson (brother of Lord Kelvin, the celebrated Scottish scientist). The developments necessary to produce a practical calculating machine were carried out in America by H. W. Nieman and Vannevar Bush, who produced the first full-scale Differential Analyser in 1930. Professor D. R. Hartree of Manchester University brought the ideas back to England, and by 1939 four large Differential Analysers had been built in this country – (at Manchester University, Cambridge University, Queen's University Belfast, and the Royal Aircraft Establishment, Farnborough). The great contribution of Differential Analysers to science and engineering was the speed at which they could solve differential

Fig. 2.3 A large differential analyser, designed by D. R. Hartree in 1935. The prototype for this was a small-scale machine built from pieces of children's Meccano construction sets, which actually solved useful equations concerned with atomic theory in 1934.

equations, compared with other contemporary methods.

Although by 1939 most of the calculating equipment available was derived from foreign designs, Britain was well to the fore in the development of computational methods. In particular, the mathematician Dr L. J. Comrie founded Scientific Computing Service Ltd in London in 1937, a company which built up a considerable reputation in the development of new techniques and the production of mathematical and astronomical tables. The Scientific Computing Service's 'human computers' (i.e. mathematical clerks) used a variety of electro-mechanical hand calculators and punched-card equipment. They offered a unique consulting service to the scientific community.

As was indicated earlier, all the calculating devices of the 1930s differed fundamentally from today's computers. Although the true stored-program computer did not become a reality until after the war, two small developments in the late 1930s can now be seen as tentative pointers in the right direction.[4] One such pointer was the project undertaken at Iowa State College between 1937 and 1941 by J. V. Atanasoff and C. Berry, who planned an electronic binary digital 'equation solver' (which was never in fact completed). The other was the series of electro-mechanical binary digital 'equation solvers' built in Berlin by Konrad Zuse, of which the first fully operational model (Z3) was working in 1941. Both projects involved a certain amount of internal storage for numbers, but not for instructions. The machines were of a type called sequence-controlled calculators, representing an important stepping-stone towards the modern computer. The exigencies of war caused setbacks to both projects. However, Zuse, with great personal tenacity and insight, started building machines again in 1948; in 1950 he founded his own company Zuse KG.

Fig. 2.4 The wheel-and-disc integrating mechanism of a differential analyser, based on an idea of the Scottish scientist James Thomson in 1876. The idea was developed further in America in the 1920s, thereby giving a means for the rapid solution of differential equations. This stimulated interest in 'automatic computation'.

3
THE SECOND WORLD WAR

The second world war brought a host of computational problems which had to be solved in a hurry. The areas for which special-purpose computing devices were designed and built included gunnery control, flight simulation and aircrew training, radar signal processing (e.g. in airborne interception) and code deciphering. The devices produced were, in general, too specialised to have any direct influence on subsequent stored-program computer development, except in one important sense. They created at places such as the Telecommunications Research Establishment (TRE) and the Post Office Research Station, groups of engineers who became very competent at designing ingenious electronic equipment. (TRE is now the Royal Signals and Radar Establishment, Great Malvern, Worcestershire.) British inventiveness was in full flood and many post-war stored-program computer projects owed a lot to the engineering impetus generated during the war years.

There is one particular wartime project which is of considerable historical interest, since it involved the first large-scale use of thermionic valves for digital computation. This was the special-purpose 'computer' named COLOSSUS, produced in great secrecy by the Post Office Research Station for the Government Code and Cipher School, Bletchley Park, Buckinghamshire. The full story of British code-breaking is still shrouded in the Official Secrets Act, but the following may be inferred from the few facts which are available (see, for example, references 5 and 6).

The German forces are known to have used two classes of machine for encoding signals prior to transmission: the Enigma series and the Geheimschreiber system. Both machines 'scrambled' the letters of a message by a complicated, virtually non-repeating mechanism of stepping

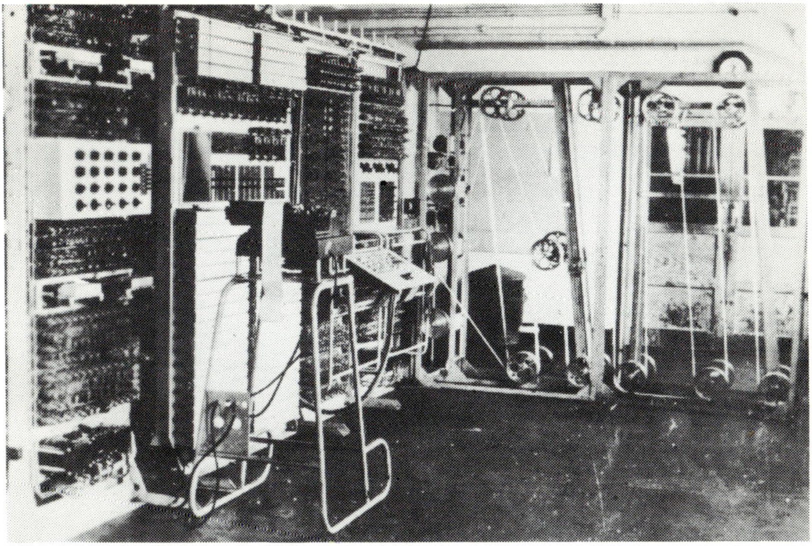

Fig. 3.1 The first COLOSSUS code-cracking machine, operational at the British Code and Cipher School at Bletchley Park in December 1943. This special-purpose 'computer' contained 1500 thermionic valves and proved that large numbers of electronic circuits could be made to do reliable calcalutions at speed. The system of pulleys to the right of the photograph guided the punched paper tape, containing the message to be decoded, through a photo-electric sensing unit. Information about trial cipher keys was generated by COLOSSUS internally during the decoding process. Partial results and the final decoded message were automatically printed out by the electric typewriter, seen to the left of centre.

rotors which rendered the encoded signal extremely difficult to decipher – unless of course the recipient had a similar machine to the sender's, adjusted to a similar set of initial rotor positions. The difference between the two systems was partly one of operational convenience: Enigma machines had a three-rotor system (four for the German Navy), and were portable; Geheimschreibers had ten rotors and used the standard 5-bit teleprinter code (Baudot code) for actual transmission. ('Bit' is an abbreviation of 'binary digit', see Appendix 1.) Enigma codes were used for all day-to-day military signals and Geheimschreibers, being regarded as more or less unbreakable, for top secret strategic messages.

There were two aspects to cracking Enigma and Geheimschreiber

Fig. 3.2 Two WRNS operators at the controls of COLOSSUS. The deciphering process was interactive, in the sense that an operator was able to alter the sequence of trial cipher keys in the light of meaningful transliterations which might appear on the machine's printer. Full details of COLOSSUS and other wartime code-cracking machines are still secret.

codes. One was to try and keep abreast of the developments which took place from time to time in the machines, for example in rotor wiring and plugboard connections and in method of use; the other was to work out the system of initial rotor settings, which were changed three times a day according to tables printed in German code-books. Inspired by pre-war code-cracking activity in Poland, the basic method used by the British at Bletchley Park for decoding was to build 'trial and error' machines which scanned quickly through many combinations of symbols until sensible transliterations were obtained. The decoding machines were, in part, Enigmas in reverse. Since three Enigma rotors had approximately 10^{21} (one thousand billion billion) possible initial states, considerable ingenuity was required at Bletchley Park to reduce the number of trial combinations to a manageable size for each message. It was only the cleverness of Bletchley mathematicians such as A. M. Turing that made decoding of large numbers of messages a practical proposition. Even so, many tens of electro-mechanical relay machines and many hundreds of WRNS operators were employed full-time on deciphering Enigma signals. Their success contributed very significantly to the allied war effort.

Deciphering the Geheimschreiber codes was much more difficult and relay-based machines proved too slow. Accordingly, in January 1943 a small team at the Post Office Research Station, led by T. H. Flowers and working to functional requirements drawn up at Bletchley Park by the mathematician M. H. A. Newman, designed an all-electronic deciphering 'computer', the COLOSSUS. This was assembled at the Post Office Research Station at Dollis Hill, London, and began useful work at Bletchley Park in December 1943. COLOSSUS contained 1500 thermionic valves – far more than any single electronic device built up to that time. The message to be deciphered was fed into an optical reader as a repetitive loop of punched paper tape in 5-bit teleprinter code, at the amazing rate of 5000 characters per second. (It was said that the hardened steel guide-pins had to be replaced frequently, because paper tape travelling at such a high speed soon wore grooves in them!) Internally, COLOSSUS contained electronics for counting, comparison, simple binary arithmetic and logical operations. Output was via an electric typewriter. The 'program', or strategy for altering trial cipher keys, was controlled from plug-boards and switches. It is believed that the deciphering was interactive, the cryptanalyst making adjustments to the program switches in the light of meaningful letter-combinations which might appear on the printer.

In 1944 a second version, COLOSSUS Mark II, was designed to keep pace with developments in the Geheimschreiber encoding rotors. COLOSSUS Mark II had conditional branching (defined in Appendix 1), was faster than its predecessors, and contained 2500 valves. The first of ten Mark II machines worked on 1 June 1944.

It is believed that COLOSSUS II contained all the elements of a modern computer *except* an internal program store. The interactive nature of the problem and the requirement for high computing speeds meant that such a store would have been an expensive and unnecessary luxury. The Bletchley Park machines were built for a specific purpose and their success in breaking the 'unbreakable' German Geheimschreiber ciphers amply justified their design. On a more general level, they proved that high-speed digital computing could be carried out reliably using thermionic valve circuits. The COLOSSUS team was aware that, once an economic storage technology became available, a universal stored-program computer would be within sight. The only trouble was that they weren't allowed to talk about their experience once hostilities ceased! On a more positive note, there is some evidence to suggest that the COLOSSUS activity did incline official thinking towards digital computers.

Before describing the post-war computing projects themselves, it is appropriate to survey the technology available at the time. We have today become so used to the convenience of the silicon chip that it is sometimes forgotten what electronic effort was required to implement even the simplest computational function in the 1940s. The next few pages also serve to introduce devices whose names have long since vanished from everyday computing vocabulary.

4
THE TECHNOLOGY OF EARLY COMPUTERS

One or two of the very early stored-program computers were *electro-mechanical*, being based on devices called *relays*. A relay is a switch which can be opened or closed automatically by appropriate electrical signals. (The activating mechanism is an electro-magnet which can 'pull' the switch contacts together.) Relays were a comparatively cheap way of implementing computing and control equipment, though they were unsuitable for use in constructing realistic storage units. Their main disadvantage for use in processors was their relative unreliability (contact-wear and susceptibility to dust) and slowness of operation. A relay took a few milliseconds (i.e. thousandths of a second) to 'switch', whereas a thermionic valve circuit could 'switch' in less than a microsecond (i.e. less than one millionth of a second). In the late 1940s computing speed, along with automatic operation, was regarded as one of the main advantages of the 'new' type of computer over its predecessors. It might thus be surmised that relays tended only to be used by research groups who could not afford anything faster!

In the period 1948–55 most computers used *thermionic valves* (or *vacuum tubes*) in their processors. There were many types of valve, giving a variety of performance characteristics. A *pentode* valve had five electrodes within it, whereas *triode* and *diode* valves had three and two electrodes respectively. A *decatron* was a special ten-state neon-based valve used in counting applications. In essence, a valve is an evacuated glass tube (sometimes inside a metal outer casing), in which some metal electrodes are made to control the flow of electrons, the electrons being produced by heating a 'cathode' electrode. In contrast, modern electronic components control the flow of charged particles inside a solid lump of semiconducting material – hence the term 'solid-state circuits'. The disadvantages of valves mainly concern their cathode heaters. These heaters consume appreciable electrical power and are, therefore, wasteful of energy. Further, the heat they produce is not all put to good use in providing electrons inside the valve, and surplus radiated heat has to be disposed of (e.g. by forced-air cooling). Finally, the heating elements deteriorate with time, especially if the equipment is being turned on and off periodically rather than left running. The early scientific reports on computers are full of statistics on numbers of valve failures per week and it is clear that component reliability was of great concern.

As far as semiconductors are concerned, the rectifying (diode-like) properties of semiconducting materials had been known about for some time. During the second world war components based on crystals of germanium came into service as the so-called 'crystal diodes'. Several early computers made use of these simple solid-state devices in conjunction with valves for logic circuits. The first major impact of solid-state components upon computer technology came in the mid-1950s when *transistors* were becoming reliable and cheap enough to replace valves altogether. The significant advantages offered by transistors were firstly ones of reduction in physical size and in power consumed. Other advantages such as reliability, cheapness and speed followed on, once the transistor manufacturing processes had evolved to a satisfactory state.

The basic properties of a transistor were discovered in 1947, at the Bell Telephone Laboratories in America. The earliest experimental use

Fig. 4.1 Early computer components. Back row left to right: an electro-mechanical relay of the 1940s; a more recent relay; a pentode valve of the 1940s in a metal outer casing; a miniature pentode of the 1950s. Foreground: a 50 pence coin for size comparison; a thermionic diode of the 1940s.

Fig. 4.2 left to right: a germanium semiconductor diode of the 1950s; the thermionic diode of figure 4.1; a silicon semiconductor diode of the 1970s.

of transistors in a computer was in 1953. Commercially available transistorised computers did not begin to appear until 1956, and it was not until the mid-1960s that valves finally became obsolescent. By 1956 the early form of *point-contact* germanium transistor was being replaced by the more reliable *surface-barrier* and *junction* types, to be followed shortly afterwards by the replacement of germanium by silicon as the basic semiconducting material for most computer transistors. The *planar* manufacturing process then replaced the 'junction' geometry in many solid-state devices, and by the end of the 1960s the planar silicon integrated circuit or 'chip' had essentially arrived. All these developments brought spectacular reductions in cost and size, whilst reliability and speed increased. The accompanying photographs illustrate these changes better than any technical description. The resulting improvements in central processor technology have given today's computer designers a freedom undreamt of by the pioneers of the 1940s.

Fig. 4.3 Flip-flops through the ages. The three units in the photograph all do approximately the same job within a computer, but represent the state of technology in about 1951, 1961 and 1971 respectively. The large 'door' is typical of early thermionic valve computers; the printed circuit board in Alison's right hand uses germanium junction transistors; the smaller board in her left hand contains small-scale silicon integrated circuit chips. The units, which all consist of flip-flop registers, come respectively from a Ferranti Mark I, a Ferranti ATLAS, and the Manchester MU5 computer. The units represent a speed improvement of about 100 times, at a decrease in contemporary cost of about 20 times.

As far as computer storage is concerned, the technological changes over the years have produced equally dramatic results, though the evolutionary path is not so clear. This is because computer storage covers a variety of ingenious devices, witness to the fact that the search for a suitable memory unit was the prime preoccupation of many early research groups. It was, of course, possible to construct a storage unit entirely from thermionic valves, using two valves per stored binary digit – the so-called *bistable* or *flip-flop* circuit. However, this was far too expensive a method, except for the limited application of temporary storage within the central processing unit itself. There were two types of high-speed storage in common use in early computers: delay lines and electrostatic tubes.

In a *delay-line* store, information was held as a train of impulses continuously circulating round a special closed path. The time taken for an impulse to circulate once round the special path was arranged to be very much longer than the time taken by electrical impulses to travel round the wires in the rest of the computer. One way of arranging the suitable 'long path time' was to take advantage of the fact that sound waves travel very much more slowly than electrical waves, and that it is possible to convert acoustic impulses into electrical impulses and vice versa. An *acoustic delay-line store* therefore had information travelling as acoustic pulses between a transmitter and receiver placed at either end

Fig. 4.4 Storage (or memory) devices for early computers. In the background is a cathode ray tube used in the Williams electrostatic storage system. The long rods in the centre of the photograph are filled with mercury and form an acoustic delay line unit from a DEUCE computer. The plug-in board in the foreground contains a coil of nickel wire for a magneto-strictive delay line in a PEGASUS computer. The small integrated circuit module to the left of centre contains a modern silicon chip which could store more information than the other three units combined.

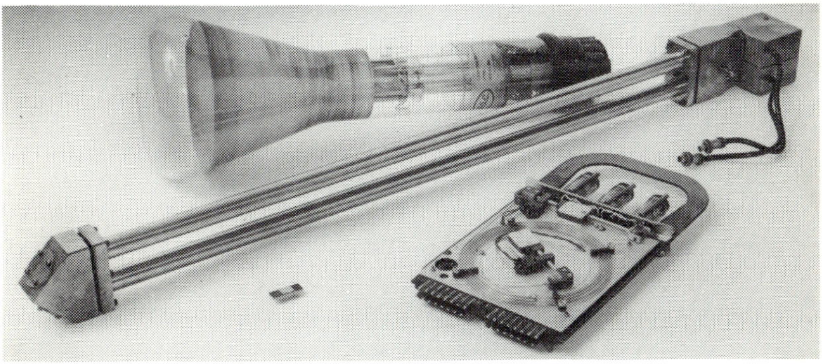

of an acoustic path. The path itself had to have suitable physical properties, a column of mercury frequently being used. For example, in the Cambridge EDSAC computer each delay line consisted of a 5 ft. steel tube filled with mercury and terminated at each end by a quartz crystal. Electrical pulses representing digits were applied to one crystal which converted them into ultrasonic waves. These travelled 'slowly' through the mercury to be picked up by the second crystal, amplified and reshaped, and sent back to the first crystal for recirculation. Close temperature control of the mercury tank was important, in order to keep the delaying properties of the system constant. A *magnetostrictive delay-line* store was also acoustic in nature, but the resulting unit was cheaper, smaller and more robust than the mercury variety. In the magnetostrictive delay line, electrical pulses were converted into stress waves which travelled down a length of (nickel) wire. The conversion from electrical pulses to stress wave and vice versa depends upon the fact that the sudden application of a magnetic field to a length of wire causes the wire to change dimension. Thus, coils similar to those found in an electro-magnet were used for inserting and recovering digital information from the delay line. Although the magnetostrictive principle was widely known, the first computers to employ it for storage purposes were those produced by the firm of Elliott Brothers in England, as described in Chapter 11.

Another class of early storage device depended on the ability of certain materials, notably phosphor, to retain an *electrostatic charge* for some time. Information was therefore stored as a pattern of charged-up areas on, for example, a phosphor cathode ray tube screen which had been selectively bombarded by a beam of negatively-charged electrons. (The phenomenon of charge-storage in phosphor is also utilised in 'screen persistence', which makes flicker-free television pictures possible.) The difficulty with electrostatic storage devices was that the stored charge leaked away – i.e. was neutralised – unless steps were taken either to retain it or refresh it. Between 1945 and 1950, various inventions, including the *selectron, barrier grid* and *holding-beam* tubes, attempted to prevent charge leakage. Another approach, used successfully in the *Williams tube*, was to refresh or regenerate the charge pattern periodically before it had time to leak away significantly. In about September 1946

F. C. Williams discovered the 'anticipation pulse' effect,[7] which made charge-regeneration, and hence long-term storage, a relatively simple matter. In particular, the timing of the anticipation pulse gave an early warning that the scanning electron beam was about to arrive at an area of charged phosphor, and the shape of the pulse determined whether this area was currently storing a binary 1 or binary 0; appropriate regeneration could then be arranged in time. Williams cathode ray tube stores were by far the most widely used form of electrostatic storage.

Fig. 4.5 Binary information stored as a pattern of bright or dim dots on the phosphor-coated screen of a Williams tube. Each dot corresponds to an area of electrostatic charge which has coincidentally been made visible by much the same property that produces areas of brightness on a television screen. Although the emission of light was not inherently part of the storage mechanism, it did mean that a Williams tube could give a primitive form of visual display for special-effect computer outputs. The actual display in the photograph shows 32 40-digit binary numbers in a Ferranti Mark I, with an extra 20-digit identification line or 'page address'. (This page address was the germ of an idea which later led to the invention of virtual addressing.)

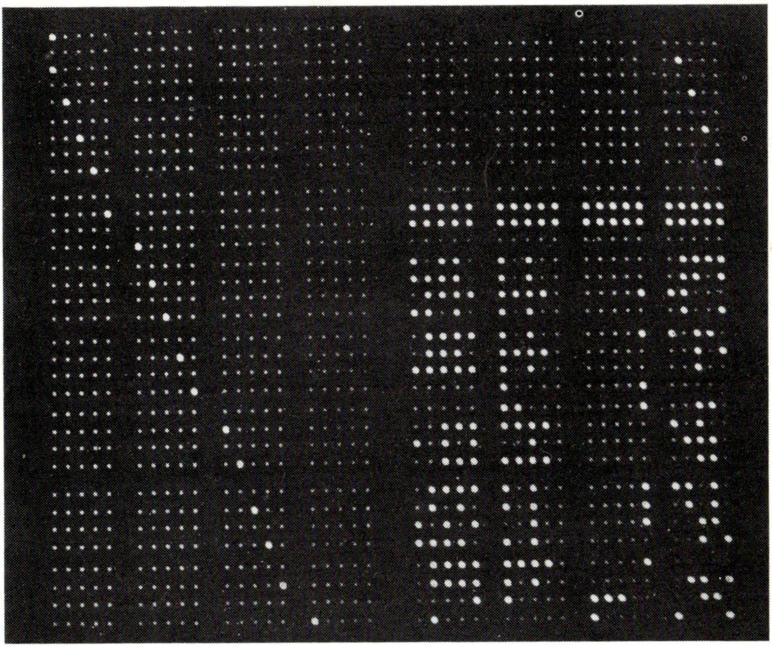

Fig. 4.6 A magnetic drum storage unit from a DEUCE computer (1955). The magnetically coated drum – the dark cylinder in the centre of the container – spins rapidly and information is recorded on its surface by read/write heads at either side. In this particular system the head-structures could be moved vertically to give the effect of many recording tracks.

Although delay-line storage was more popular for use in early British computers, Williams tubes narrowly outweighed delay lines as a choice for storage device in early American computers.[8] An extremely desirable property of a fast store is that it should be *random access* – that is, the time to access information at any one location should be the same as the time taken to access any other location. This 'independence of location' is very important when accessing programs, because of the out-of-sequence branches (control transfers) which they contain. A disadvantage of early delay-line stores was that they were *sequential* access devices, though this inconvenience could be minimised by a combination of actions taken at the hardware and software level. Williams tubes were random-access devices, and were also more suitable for the design of parallel (as opposed to serial) computers.

A third class of computer storage device, still widely used, is based on the ability of certain materials to retain (i.e. store) their state of magnetism. Thus magnetic *tapes*, *films* (complete with sprocket holes), *discs* and *drums* all basically record information on a magnetic surface ready for subsequent 'playback'. ('Writing' and 'reading' are the terms normally used in place of 'recording' and 'playback', but the principle is much the same as that used in an ordinary audio cassette tape recorder.) Storage units based on magnetic surface recording inevitably depend for their access-time on the speed at which the read/write head moves past the surface. This speed has always proved much too slow for the potential rate of information flow demanded by a computer's central processor. Thus in some early machines, as in most present-day ones, magnetic surface recording is useful mainly as a back-up to a faster unit. In this role its advantage lies in its comparatively low cost per stored bit. This economic relationship is of course relative. In absolute terms the actual capacity (quantity of information) contained in both the fast and the slow section of a computer's store has typically increased at least a thousandfold between 1950 and 1980.

Historically, the demise of delay lines and electrostatic tubes for fast storage units came between about 1954 and 1958 with the introduction of the *magnetic core store*, an important American invention associated with J. W. Forrester of the Massachusetts Institute of Technology. In this store, each binary digit is remembered as the magnetic state of an individual ring or 'core' of ferrite material. An array of many such cores, suitably threaded by selection and sensing wires, goes to form the complete *core store*. These stores had access times much closer to those required by central processors. For general-purpose computers, core stores lasted well into the 1970s. Nowadays, stores based on semiconductor integrated circuits have replaced cores as the common form of fast storage device, except in certain military applications where cores are preferred for their relative insensitivity to power supply interruptions.

Having gained an idea of the electronic techniques in the 1940s, let us return to the trail of the first proper stored-program computer.

Fig. 4.7 A 'stack' of magnetic cores, compared in size with a match box. This particular stack is made from 48 planes, each plane wired with a mesh of 32×32 tiny rings (or cores) of ferrite material, enabling the complete unit to store 1024 48-digit binary numbers. This core store dates from about 1959.

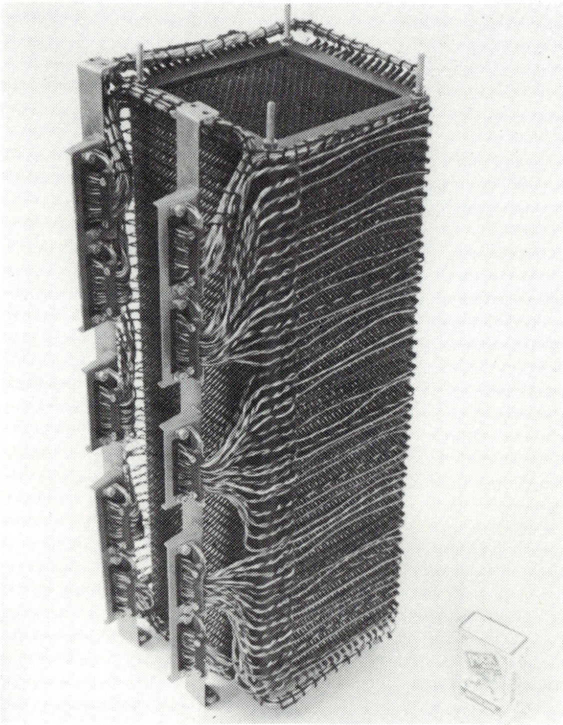

5
THE ACE, THE 'BRITISH NATIONAL COMPUTER'

At the close of the second world war there was some feeling amongst scientists at the Ministry of Supply that a National Mathematical Laboratory should be established to coordinate facilities and techniques relating to machine-aided computation. The practical outcome of this feeling was the establishment in the summer of 1945, of a Mathematics Division in the National Physical Laboratory (NPL) at Teddington, Middlesex. Amongst the father-figures associated with this event were Professor D. R. Hartree of Manchester University and (from October 1946) Cambridge University, and Dr L. J. Comrie, founder of the Scientific Computing Service in London. Hartree was the chief link between British and American computing efforts in the immediate post-war years; both he and Comrie had accumulated a great deal of experience on mechanical and electro-mechanical calculators.

As far as *building* a stored-program computer, the initial enthusiasm came largely from a group of people who had been involved with the COLOSSUS deciphering activity at Bletchley Park. In particular, amongst the Bletchley team which disbanded in the autumn of 1945 were the mathematicians Professor Max Newman and Dr Alan Turing, and the Post Office engineers T. H. Flowers and Dr A. W. M. Coombs. In October 1945 Newman moved to Manchester University, where he wished to set up a 'calculating machine laboratory'.[9] His plan was to construct a stored-program computer similar to one being proposed by John von Neumann of Princeton University, using a special storage device called the Selectron tube. As is explained in Chapter 16, von Neumann's group at Princeton grew out of the American ENIAC development at the Moore School of Electrical Engineering, University of Pennsylvania, Philadelphia. This group was the source of one of the

Fig. 5.1 Map showing the location of some centres of early computer activity. The National Physical Laboratory is at Teddington, the Telecommunications Research Establishment at Malvern, and the wartime Code and Cipher School at Bletchley.

first formal proposals for a stored-program computer – the 'EDVAC report'.[10] The Selectron tube, under development by the Radio Corporation of America from about 1945, was for a time thought to be the most promising digital storage device. Max Newman's plans at Manchester were in the end overtaken by events: the Selectron ran into prolonged technical difficulties and the Princeton computer was not working until 1952, with a different storage mechanism; meanwhile a completely independent computer had been built by the Electrical

Engineering Department at Manchester – as described in Chapter 7. (In passing it should be said that, despite the relatively late completion of the Princeton project, there is little doubt that John von Neumann himself was the most influential of all the early computer pioneers.)

Of the other Bletchley personnel, Turing joined the new Mathematics Division at NPL where, with characteristic energy, he immediately set about designing a universal computer. Turing had written an important theoretical paper on computers in 1936[11] and, although familiar through personal contact with von Neumann, he had no need or inclination to copy anyone else's design. On 19 February 1946 he presented to the Executive Committee of NPL, probably the first complete design for an electronic stored-program computer,[12, 13] including a cost estimate of £11,200. It is likely that Sir Charles Darwin, the NPL Director, thought of Turing's proposal for an Automatic Computing Engine (ACE) in terms of a single national effort which would result in a computer housed at NPL and serving the needs of the whole country. At that time the NPL establishment at Teddington did not readily have the capability for constructing a large national computer and Darwin was therefore anxious to enlist the cooperation under contract of other relevant organisations such as the Post Office.

It is clear that at the Post Office's Dollis Hill research laboratory both Flowers and Coombs, ex-Bletchley men, had the necessary interest and expertise to implement a stored-program computer, and indeed the immediate post-war intention was for the Post Office to do just that. Flowers had visited the Moore School, University of Pennsylvania – the main centre for US computer activity – late in 1945. He has since said, 'Unfortunately the pressure of telephone reconstruction after the war left so little effort for other projects that eventually the commitment [to build a computer] had to be withdrawn. Some mercury delay lines were constructed but little else.' These particular lines resulted from a development contract placed by NPL in June 1946; a prototype 1000-bit delay line was working at Dollis Hill by 21 January 1947.[14] Another Post Office commitment, to the Telecommunications Research Establishment (TRE) at Malvern, involved the construction of a computer called MOSAIC for defence applications; this spanned the period 1947–54 and is described in Chapter 10.

Having failed to secure the help of the Post Office in building a national computer, Darwin and the NPL Mathematics Division turned their attention in the autumn of 1946 to two new British groups which had emerged as likely sources of computer electronics expertise. The first was a small team led by Dr (later Professor Sir) F. C. Williams at TRE; the second was a group led by Mr (later Professor) Maurice Wilkes at Cambridge University. Williams had been working since July 1946 on a novel binary storage system using conventional cathode ray tubes. Wilkes had attended a series of lectures on the American EDVAC proposal for a stored-progam computer, held in August 1946 at the Moore School, University of Pennsylvania, and had returned determined to build a similar machine at Cambridge.

As a result of Darwin's approach to TRE a party led by Dr R. A. Smith (superintendent, TRE Physics Division) and including Williams and Dr A. M. Uttley, visited NPL on 22 November 1946. A report of this meeting survives[15] and is worth quoting from because it illustrates

Fig. 5.2 Some British computer pioneers and other distinguished guests at the opening of the Science Museum's computing gallery in London, December 1975. Back row, left to right: Donald Davies, Tommy Flowers, Grace Hopper (USA), Jim Wilkinson, Tom Kilburn, Raymond Thompson, Maurice Wilkes, Cecil Marks, Allen Coombs. Front row: Mrs Douglas Hartree, Freddie Williams, Max Newman, David Wheeler, Konrad Zuse (Germany).

the scarcity of expertise at that time:

> The director [Sir Charles Darwin] emphasised the extreme importance which he attached to the development of A.C.E. and put it as having the highest priority in his opinion of any work that was being done for D.S.I.R. at T.R.E. [D.S.I.R. was the Department of Scientific and Industrial Research, a predecessor of the present Science Research Council.] He was most anxious that some effort should be set aside for work on this project. Dr Smith explained the difficulties in which T.R.E. found itself at the moment as regards staff knowledgeable in electronic circuit technique. Apart from the small number of staff now working for Dr F. C. Williams most of the able circuit technicians had been transferred to the Department of Atomic Energy. Dr F. C. Williams had himself been appointed to a professorship in the University of Manchester and was leaving in six weeks. Dr A. M. Uttley had staff knowledgeable in computing technique but not expert in valve circuits. These staff were, however, almost completely tied up with important work for the Ministry of Supply on computers for military applications and it was unlikely that much effort would be available from that source. It therefore seemed likely that the only way in which T.R.E. could continue to contribute would be to second a small number of staff, say one Scientific Officer and an assistant, to work under Dr Williams' direction at Manchester University, and for a small team to be constructed from the remainder of his staff, and possibly one or more from Dr Uttley's present staff, under the direction of Dr Uttley.
>
> The purpose of the larger meeting was mainly directed towards elucidating the present position of the development of the A.C.E. It appeared that although an elaborate paper design had been laid down, the fundamental problem of storage of information has not been solved and that, as had been suspected, the experimental work of Dr Williams's at T.R.E. on storing information on a cathode ray tube was considerably in advance of the work which the Post Office were doing on the use of delay lines for storage purposes. There was therefore necessary a considerable amount of basic investigation on storage systems before the computer could actually be brought to the

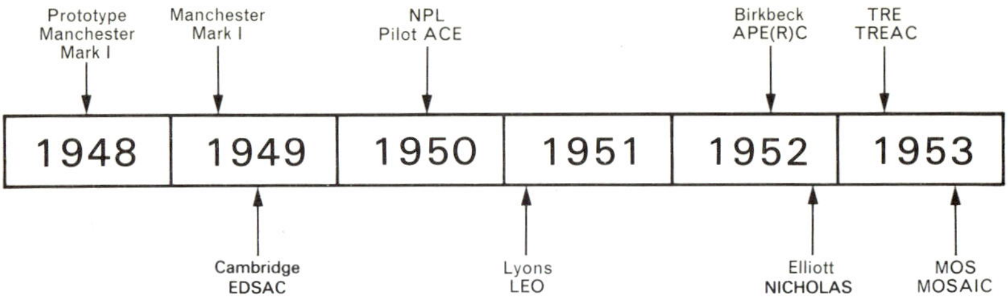

Fig. 5.3 Time-chart showing the date on which some of the vintage British stored-program computers became operational. Details of these and many other machines are given in later chapters.

> stage of being assembled. Switching techniques were also discussed and it appeared that although the knowledge existed for making electronic switching systems, a good deal of work would have to be done before suitable practical designs were evolved. This knowledge, however, mainly existed with Dr F. C. Williams and with the group in T.R.E. engaged on the development of counters for the Department of Atomic Energy.

Dr Uttley's group at TRE went on to build TREAC,[16] a parallel computer using Williams tube storage and completed by mid-1953 (see Chapter 10). Williams, who had an outstanding reputation as an inventor, left TRE for Manchester University in December 1946. He took with him Tom Kilburn, who was later to become the driving force behind all subsequent Manchester computer development. As for helping with the ACE project, an NPL proposal and draft contract was sent to Manchester in December 1946 but was declined shortly afterwards by Williams. The reasons for Williams' lack of interest were threefold:

(*a*) there was at that time no need for Williams to seek additional financial support for his own work;
(*b*) the wording of the NPL draft contract appeared restrictive compared with normal academic practice; and
(*c*) Turing's ACE design, and the personality of Turing himself, were incompatible with Williams' own way of pursuing research.

NPL's interaction with Cambridge University was spread over a somewhat longer period, beginning in the autumn of 1945 when Wilkes had been released from wartime work at TRE to take up the Directorship of the newly formed University Mathematical Laboratory. An impression of the initial Cambridge activity may be had from the following quotation from a letter by Wilkes to J. R. Womersley, Superintendent of NPL Mathematics Division, dated 2 April 1946:

> We are just getting organised and I want to decide what our research programme on new calculating machines shall be. I would like to have a discussion with you so that anything we do may be coordinated with your own activities. You will probably have suggestions to make as to how we can best apply our limited efforts.
>
> I am now rather more in the picture as to American activities than when I last discussed the subject with you. Professor Hartree was good enough to write to Philadelphia and get permission to tell me officially about the American projects and he gave me a copy of the report he wrote after his last visit to the States.

Discussions concerning a formal contract between Cambridge and NPL gained momentum in November 1946, by which time Wilkes' group had initiated the building of what was to become the Cambridge EDSAC stored-program computer. By April 1947 the idea of an NPL/Cambridge contract had been dropped, for much the same reasons as had applied in the case of Williams. Wilkes went on to complete EDSAC by May 1949 (see Chapter 6).

All these negotiations left Darwin and the concept of a British national computer somewhat at a loss. A new Electronics Division was set up at NPL in August 1947 and it was here that the Pilot Model of the ACE computer was built, eventually running its first program on 10 May 1950; this development is described in Chapter 8. The summer of 1947 also marked Turing's effective departure from NPL. He spent a sabbatical year at his old Cambridge College and then moved, in the autumn of 1948, to the Mathematics Department at Manchester University, run by his friend Max Newman. Turing is remembered as a truly brilliant mathematician who many found very difficult to work with. It is significant that his original ACE proposal was radically different from any other contemporary computer design and that, although spending time at Cambridge and Manchester, he did not materially contribute to the design of either of those Universities' machines. At Manchester he eventually devoted most of his energies to developing early programming techniques and solving partial differential equations on the computer in connection with his theory of morphogenesis, the 'growth and form of living things'. He died suddenly in June 1954.

Any possibility of a 'British national computer project' had evaporated by 1947. If personalities and resources had been otherwise, it is quite likely that such a computer would have been completed by the Post Office or NPL, acting independently or jointly, by 1948. Even if this had occurred, it is however most unlikely that the austere post-war economic climate would have permitted any immediate industrial exploitation. It should be remembered that wartime petrol rationing did not end until May 1950, and food rationing of one form or another lasted until July 1954. There were problems enough without the scientific oddity of a universal computer!

In order to preserve the chronological sense of the story, we must now leave NPL pondering Turing's ACE design, and move the spotlight to the Universities.

6
THE CAMBRIDGE EDSAC

Maurice Wilkes' Cambridge University group was perhaps the most coordinated computer design team anywhere in 1947. It included W. Renwick, S. Barton and G. Stevens on the hardware side and later D. J. Wheeler on the programming side. Their objective was to set up a usable and reliable computing service (for university research workers) in a short timescale; they were therefore not necessarily interested in

Fig. 6.1 Part of the Cambridge University EDSAC computer during construction in 1947. The photograph shows the following members of the design team (left to right): G. J. Stevens, J. Bennett, S. A. Barton, P. Farmer, M. V. Wilkes, W. Renwick, R. Piggott. Professor Wilkes is kneeling beside a mercury delay line assembly.

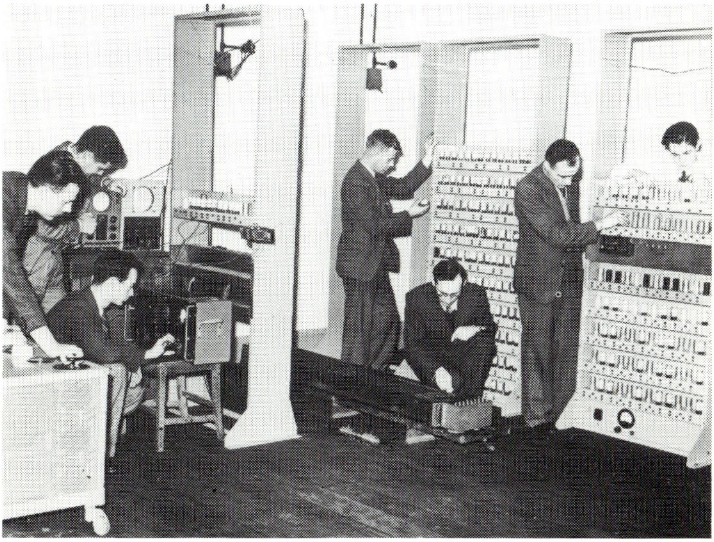

Fig. 6.2 The completed Cambridge EDSAC in 1949. This famous computer quickly settled down to provide an efficient computing service and helped to put Britain well ahead in the implementation of practical computers and their software.

building the 'best possible machine'.[17] Their stored-program computer became known as EDSAC (Electronic Delay Storage Automatic Calculator) and, as the nomenclature implies, its design was influenced by the American proposals[10] for a machine called EDVAC (Electronic Discrete Variable Automatic Calculator).

The development of a suitable storage unit was the principal problem facing all the early computer designers. It was mainly because of storage difficulties that the American EDVAC did not see the light of day until 1952. In Wilkes' case he chose mercury delay lines for EDSAC because encouraging results had been achieved with this technique towards the end of the war by the Admiralty and others, for storing pulses in a device designed to improve the tactical clarity of radar displays. It was fortunate that a research physicist at Cambridge, T. Gold, had worked on delay lines at the Admiralty Signals Establishment and was able to provide the EDSAC team with accurate constructional details.[18] In EDSAC each mercury-filled tube or 'tank' was about 5 ft long and stored 576 binary digits. The main store consisted of 32 such tubes, with additional tubes acting as central registers within the processor.[19]

Fig. 6.3 The direct commercial descendant of the Cambridge EDSAC was a computer named LEO (Lyons Electronic Office) shown here in about 1953. LEO pioneered the use of computers for business data processing, as described further in Chapter 13.

A full technical description of EDSAC is given in Appendix 2, together with a comparison between it and other contemporary computers. Briefly, EDSAC stored 512 36-bit words (a 'word' being a general term for a collection of bits treated as a whole by the computer). The time taken to perform an addition instruction was 1.4 milliseconds. Input was via a 5-bit electro-mechanical paper tape reader and output was to a teleprinter. EDSAC contained about 3000 thermionic valves and filled a large room. The financing of the project was via normal University research channels, plus a (for those days) sizeable donation of £2500 from J. Lyons & Co. Ltd. (The connection between this company and Cambridge is an interesting one, to be described later.)

EDSAC first ran a program on 6 May 1949, and offered a regular computing service from early 1950 until its shut-down in July 1958. This was the first such service in the world to be run on a proper stored-program computer. The EDSAC group wished to make life easy for the inexperienced user, so programming was in terms of an elementary *symbolic* assembly language, whereby the programmer wrote out his instructions in terms of meaningful alphabetic characters. These were punched on paper tape, read into EDSAC, and then converted automatically to the basic binary machine instructions. An example of the resulting convenience to the user is given in Chapter 15, where four early programming systems are compared. A significant feature of the Cambridge computing service was the availability of a well-stocked library of standard subroutines held on paper tape. Whilst not the only advocates of subroutine libraries, the Cambridge group were nevertheless very influential in setting standards for their use, and in pioneering programming techniques in general. The group's book *The preparation of programmes for an electronic digital computer* (Wilkes, Wheeler and Gill, 1951) was the first textbook on programming a stored-program computer. Stan Gill was at Cambridge from 1949 to 1954.

The active encouragement of computer development in Britain was not only a characteristic of Douglas Hartree but also of Maurice Wilkes and the Cambridge Mathematical Laboratory as a whole. A major international computer conference was organised at Cambridge in June 1949, and a regular series of computing seminars was later initiated by the Mathematical Laboratory. These well-remembered seminars ensured that almost all the British computer teams were in frequent contact with each other. Wilkes and the EDSAC group were thus of considerable inspiration to other pioneers, as well as being outstanding pioneers in their own right. Knowing they had something worth selling, they even made a film of how to use EDSAC – surely a world first! Stills from this epic may be seen in reference 18.

On a more general front, in 1951 Wilkes was the first to expound clearly the principle of microprogram control as a basis for machine design (see Appendix 1, Section C). Microprogramming has had a major influence on the implementation of small and medium-range computers.

In addition to the work on programming and subroutine libraries, an interest in computer design was maintained at Cambridge after 1949. Engineering modifications and enhancements to EDSAC were carried out from time to time until the mid-1950s. A new computer called EDSAC II was then conceived, this becoming operational in 1957. Some technical details of EDSAC II are given in Appendix 2, from which it can be seen that this machine had a very respectable turn of speed.

Regarding hardware design, the main influence of the original EDSAC on other projects lay in the close informal association with the J. Lyons catering company. Early in 1947, T. R. Thompson of Lyons had begun to investigate the possibility of designing computers for use in offices. Contact between J. Lyons & Co. and Wilkes' team was first suggested by Dr H. H. Goldstine (of Princeton), and the cooperation became a reality in the autumn of 1947. From 1949 onwards, T. R. Thompson and J. M. M. Pinkerton supervised the construction of the LEO (Lyons Electronic Office), which was a re-engineered version of EDSAC with the same instruction set. Construction of LEO units was contracted out to other firms such as Wayne Kerr Laboratories Ltd, and the Coventry Gauge and Tool Co. The 'LEO Chronicle' kept by T. R. Thompson indicates that LEO was able to run simple test programs in the spring of 1951 and was doing regular clerical jobs for J. Lyons by November of the same year. Armed with an entirely new input/output system which at first included the provision of magnetic tape storage, LEO was performing a full business data processing service by the end of 1953. In addition, various scientific jobs were being undertaken for outside users, such as the Ministry of Supply and de Havilland Propellers Ltd. By mid-1954, Pinkerton was submitting proposals for a LEO II and on 4 November 1954 a new company, LEO Computers Ltd, was founded. The wider implications of these industrial developments are mentioned again in Chapter 13. Meanwhile we should look at an independent line of research which was being pursued in Manchester.

7

THE MANCHESTER MARK I

As has been noted, F. C. Williams and Tom Kilburn arrived at Manchester University in December 1946, with the intention of developing a novel form of computer storage using conventional cathode ray tubes (CRTs). This became known as the Williams tube. Bits of information were actually stored as very small areas of electronic charge, put on the phosphor-coated screen of the CRT by a controlled beam of electrons. Since the charge would leak away in about a fifth of a second, the pattern of information had to be continually refreshed. The success of Williams tubes was based upon the discovery of a relatively simple method for effecting this regeneration. Williams tubes could, moreover, be built comparatively cheaply from standard components and the ideas were soon taken up by other computer groups in the USA and elsewhere. The significant advantage of this memory over other contemporary storage systems was that Williams tubes allowed *random access* to word locations, as opposed to the *sequential* access mechanism inherent in delay-line stores.

Back at Manchester, 1947 was spent perfecting the storage system, Kilburn publishing the results together with the outline design for a hypothetical computer in December of that year.[7] The team was joined by G. C. Tootill from TRE, and a very small prototype computer was then built round a 32×32-bit word Williams tube store, in order to subject the new memory system to the 'most searching tests possible'. Though small, this machine contained all the elements of a stored-program computer but with manual input from a keyboard and output to a monitoring display screen. This prototype ran a 52 minute factoring program on the morning of 21 June 1948,[9] thus becoming the world's first stored-program computer to operate.

Encouraged by this success, the Manchester group doubled its size in September 1948 by taking on two research students (D. B. G. Edwards and G. E. Thomas), and began to expand the computer into a useful facility. An initial stimulus arose from some problems in number theory which had been suggested by Professor Max Newman. The first realistic problem to be solved, an investigation into Mersenne prime numbers, was run in early April 1949, by which time several improvements had been added to the machine. In fact the engineers' enthusiasm for improvements tended to outweigh any call for a stable computing environment, and it is significant that 42 computer patents emanated from Manchester during the period 1948–50. Although continual enhancement obviously affected reliability, an overnight error-free computing run of 9 hours was recorded on 16/17 June 1949.

In an age when acronyms were popular, the Manchester Mark I was sometimes referred to as MADM (Manchester Automatic Digital Machine) or MUC (Manchester University Computer). However, the designers usually refer to it simply as the Mark I. A technical description of the various stages of Mark I development between June 1948 and October 1949 is given in Appendix 2. In summary, the Mark I contained storage for 128 40-bit words in fast Williams tube storage, backed up by 1024 words on a slower magnetic drum store. An addition instruction was performed in 1.8 milliseconds. Input and output was via a 5-bit paper tape reader and a teleprinter. The Mark I was mostly built out of war surplus thermionic valves supplied by TRE. The cost of the project was therefore small.

Both the geographical location of Manchester and the personal inclination of Williams and Kilburn lent a certain independence to the research, with the team 'inventing things as the need arose'. In the wider context, two aspects of the Mark I design stand out. The first is the Manchester invention of *index registers*, a feature now seen on every modern computer (see Appendix 1, Section C). The second interesting aspect of the Mark I was the early combination of a small, but fast, random-access store backed by a slower (but larger capacity) sequential store. Observations on the information flow between these two levels were to produce significant Manchester inventions incorporated in the ATLAS computer, as mentioned in Chapter 9.

Fig. 7.1 Part of the Manchester University Mark I computer as extended in 1949. The prototype, operational in June 1948, occupied the centre six racks in the photograph and is believed to be the world's first stored-program computer. The Mark I was built out of war-surplus components with an enthusiasm that left little time for tidyness! Incredibly, it worked for long enough at a time for some useful computation to be carried out. It also contained some novel ideas, for example index registers.

The University Mark I was dismantled in August 1950 to make room for further computer developments. One of the last users was Dr D. G. Prinz of Ferranti Ltd, who computed Laguerre functions in connection with the control of guided weapons.

The credit for ensuring that the Mark I's light was not hidden under an academic bushel probably goes to Professor P. M. S. Blackett (later Lord Blackett), an influential scientist and government adviser. He suggested that Sir Ben Lockspeiser, the then government Chief Scientist,

should see the prototype computer whilst paying an informal visit to Blackett in October 1948. G. C. Tootill, Kilburn's colleague at the time, was able to give Lockspeiser a convincing demonstration of the Mark I which so impressed him that within a few days he initiated a government contract with the Manchester firm of Ferranti Ltd to make a production version of the machine 'to Professor Williams' specification'. (This is surely an all-time record for administrative speed and brevity!) This resulted in a five-year contract running from November 1948, involving an estimated £35 000 per annum. Perhaps more importantly, the contract established a fruitful link between Manchester University and the computer industry which has been maintained through five projects up to the present day.

Ferranti Ltd delivered nine production Mark I and Mark I Star computers between 1951 and 1957, the 'Star' developments subsequent to November 1951 receiving financial support from the National Research Development Corporation (NRDC). The first Ferranti Mark I[20] was installed at Manchester University in February 1951, thereby becoming the world's first commercially available computer to be delivered. It was identical in design to the October 1949 University machine, except in the matters of detail given in Appendix 2. The machine contained about 4050 thermionic valves and consumed about 25 kilowatts of power.

The Ferranti Mark I provided a computing capacity far in excess of the University's own needs at that time, and so outside users were actively encouraged. Most of these users came from industry or other Universities, though Dr Alec Glennie, from the government's Fort

Fig. 7.2 Some of the Manchester Mark I design team. Left to right: D. B. G. Edwards, F. C. Williams, T. Kilburn, A. A. Robinson, G. E. Thomas. Professor Williams is seen peering at the operator's monitoring and display screen. In an age when acronyms were popular, this computer was sometimes called MADM – Manchester Automatic Digital Machine.

Fig. 7.3 Symbol of a fruitful collaboration between university and industry: Brian Pollard (left) and Keith Lonsdale of Ferranti Ltd, seated at the console of a Ferranti Mark I with Tom Kilburn in the background. This machine, the production version of the University prototype, was delivered in February 1951 and is thus believed to be the world's first commercially available computer. To the right of the photograph some of the machine's panels have been removed to reveal the same type of circuit-carrying door as is shown in figure 4.3.

Halstead establishment, carried out extensive calculations for atomic weapons. (So secret were the calculations that he was required to lock himself in the computer room, and even destroy the printer inking-ribbon upon leaving!) As a hobby Glennie developed in 1952 his 'autocode', and what is believed to be the world's first compiler.[21] Glennie's system was not available to the general user, who had to wait

Fig. 7.4 The Manchester University Mark II computer, nicknamed MEG, was 20 times faster than Mark I yet consumed less than half the electrical power. It first worked in May 1954.

until March 1954 when R. A. Brooker introduced the Mark I Autocode language – a distinct improvement on the rather primitive machine-code conventions which Manchester programmers had had to endure. (These conventions, instituted by Alan Turing, required users to write their programs in terms of the alphabetic symbols of the 5-bit teleprinter code. To Turing, who had spent countless hours at Bletchley Park battling with Geheimschreiber 5-bit ciphers during the war, the teleprinter code must have seemed very natural. To lesser mortals it was painful! An example of the agony is given in Chapter 15.) As in most projects, there were moments of light relief. The following is a love letter, 'created' by the Ferranti Mark I in an unpredictable way by making use of an electronic random number generation facility which the machine possessed:

> Darling Sweetheart,
> You are my avid fellow feeling. My affection curiously clings to your passionate wish. My liking yearns to your heart. You are my wistful sympathy: my tender liking.
>
> Yours beautifully,
> M. U. C.

From 1951 the Williams/Kilburn design team was working on a Mark II computer[22] nicknamed MEG (Megacycle engine). This had a similar structure to the Mark I from the user's view, except that it avoided many of the problems associated with numbers overflowing the 'space' allocated (i.e. MEG used floating-point arithmetic). It was possibly the first stored-program computer to include this facility. Electronically, MEG was some 20 times faster than the Mark I, yet was more compact and consumed less than half the power. MEG first ran a program in May 1954. It was the prototype for the Ferranti Mercury computer, of which the first one was delivered to the Norwegian Defence Research Establishment in August 1957 as described in Chapter 14.

It is now necessary to go back in time a few years to pick up the threads of computer development at the National Physical Laboratory, Teddington. It will be remembered that the mathematician A. M. Turing produced a design proposal for the ACE stored-program computer as early as February 1946, but that the project had run into organisational difficulties. Turing had effectively left NPL in the summer of 1947.

Fig. 7.5 The Ferranti MERCURY, the production version of MEG, shows the wardrobe-like impersonality of the large modern computer – with the difference that in 1957 each programmer still had complete control of the machine. This accounts for the clock and the comfortable chair!

8
THE NPL PILOT ACE

The construction of a pilot version of Alan Turing's ACE proposal had got under way in earnest by mid-1948. The design team was drawn from both the Mathematics and Electronics Divisions at NPL. The leading lights were Dr J. H. Wilkinson (Mathematics) and E. A. Newman (Electronics). The former was an NPL colleague of Turing's; the latter had been recruited from EMI Ltd when the Electronics Division was formed.

The Pilot ACE had a complicated instruction format,[23] and was quite unlike any other contemporary machine. An impression of Turing's approach may be inferred from his barbed comment of December 1946, on being shown an outline proposal for Wilkes' Cambridge EDSAC: 'The "code" which he [Wilkes] suggests is however very contrary to the line of development here [at NPL], and much more in the American tradition of solving one's difficulties by means of much equipment rather than by thought.' (This is rather unfair on the rest of the world: Turing's design may have been economical in equipment but it certainly caused the programmer to work hard, as can be seen in Chapter 15.)

As is shown in Appendix 2, the basic internal clock frequency of the Pilot ACE was, at 1 megacycle, the fastest of the early British computers. Instruction times were highly dependent on the position in store of the instruction and an addition could take anything between 64 microseconds and 1.024 milliseconds. The main store consisted initially of 128 32-bit words in mercury delay lines. This was extended to 352 words by the end of 1951, and a 4096-word drum store was added in 1954. Since the NPL already had a large Hollerith punched-card calculator it was sensible to make cards the medium for both input to and output from the Pilot ACE. The computer contained 800 thermionic valves and some help

Fig. 8.1 Alan Turing in 1951, on being elected a Fellow of the Royal Society at the age of 39. This brilliant mathematician went too far and too fast for many of his contemporaries, thus making the long-term influence of his ideas on computer design rather less than perhaps it should have been. His highly original proposals for the Automatic Computing Engine (ACE) date from late 1945. Turing died in tragic circumstances in 1954.

with the construction of the chassis was obtained from the English Electric Company. In concept the machine was similar to Turing's original 1946 proposal, but with some interesting changes, as noted in Appendix 2.

The Pilot ACE first ran a simple program in May 1950 and was successfully demonstrated to the press in December of that year. After

Fig. 8.2 The Pilot ACE computer being demonstrated to the Press at the National Physical Laboratory in December 1950 when, to the delight of its designers, it performed with unprecedented reliability. A mixture of apprehension and relief may perhaps be detected on the faces of (left to right) G. G. Allway, E. A. Newman and J. H. Wilkinson. Much of the Pilot ACE can be seen today in the London Science Museum. (Note: DSIR stands for Department of Scientific and Industrial Research, the government body responsible for the National Physical Laboratory.)

some modification it went into full-time use as a service computer in February 1952, and it can be inferred that unreliability caused some headaches before that date. In its structure the Pilot ACE was a remarkable machine, achieving very high performance from a relatively small amount of electronics. The design, which was essentially created in 1945, is so obviously original that it belies the oft-expressed view that all stored-program computers stem from the pioneering EDVAC report. It is clear that Turing's ACE design was, in a word, unique.

Computer design continued at NPL between 1953 and 1957 on the full (final) version of ACE.[24] This had a 48-bit word, employed delay-line storage, and had a multiplication time of about 448 microseconds. It contained about 7000 valves and was working by late 1957. Meanwhile, cooperation with the English Electric Company resulted in a properly engineered version of the Pilot ACE. This was appropriately named the English Electric DEUCE. The first production DEUCE was delivered in 1955, one of the early installations being at the NPL itself in March of that year. This and other English Electric developments are described in Chapter 13. Although NPL are today continuing to make contributions to particular areas of computer science, 1957 saw the conclusion of their excursions into the field of building large computers.

The completion of ACE marked the end of one line of influence emanating from the wartime deciphering work at Bletchley Park. The other line, nurtured by T. H. Flowers and A. W. M. Coombs at the GPO's Dollis Hill research laboratory, will be described after we have first caught up with some developments in the application of transistors to computers.

9
TRANSISTOR COMPUTERS

All of the stored-program computers described so far used thermionic valves. Although the physical operation of a transistor was discovered at Bell Telephone Laboratories before the first computer actually worked, the devices were for some years only of interest to electronic research groups. This was because of the difficulties experienced in manufacturing reliable transistors of the point-contact variety then in use. Bell Telephone Laboratories naturally experimented with the use of transistors in computing circuits and this led to an Air Force contract to build a special-purpose computer called TRADIC, whose 'program' was set up manually on a plug-board. TRADIC[25] contained 700 point-contact germanium transistors and was working under test conditions in the spring of 1954.

Despite the American technological lead, it was actually in Britain that the first proper transistor computer came into operation. Concurrent with the MEG development mentioned previously, Kilburn's group at Manchester University built a small transistor research computer to gain some experience with the new devices. R. L. Grimsdale took a leading part in the design. Two versions of the computer were completed – in November 1953 and April 1955 respectively. Both versions had a one-plus-one address instruction format and a drum store; the second machine had an extended instruction set and more storage.[26] Although somewhat unreliable and slow (the average instruction time was 30 milliseconds), the November 1953 computer is believed to have been the first transistorised machine to run a program. The Metropolitan–Vickers Company adapted the prototype's design to form the basis of their MV950 computer. The first production model was completed in 1956 and six MV950s were made – mainly for internal use within the Metropolitan–Vickers organisation. The production MV950 used the

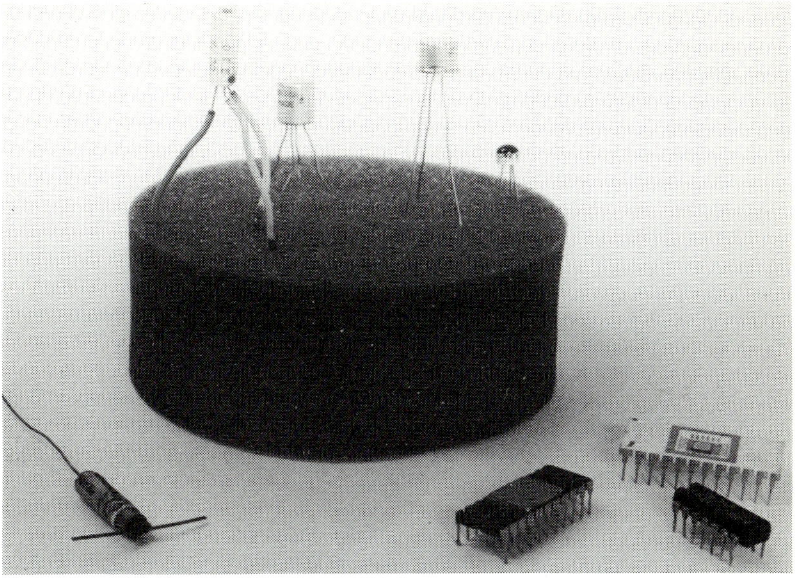

Fig. 9.1 Transistors old and new. Upper group, left to right: two germanium junction transistors of about 1957 and 1959 respectively; two silicon planar transistors of about 1963 and 1968 respectively. Lower left: a germanium point contact transistor of about 1953; lower right: a group of three silicon planar integrated circuit modules of the 1970s. The far integrated circuit module has its cover removed to reveal a silicon chip measuring about 0.5 centimetres square. This chip contains the equivalent of about 700 transistors and 1000 resistors, and is typical of large-scale integration of the late 1970s. The mysteries of all the semiconductor devices shown in the photograph derive from the original American invention of the transistor at Bell Laboratories in 1947.

more reliable junction type of transistor.

The early transistor circuit experiments at Manchester were very similar to those carried out at the same time by the UK Atomic Energy Research Establishment, Harwell. At Harwell the experiments led to the design of the CADET computer, which first ran a simple test program in February 1955.[27] From August 1956 CADET was offering a regular computing service, by which time the machine contained 324 point-contact transistors and 76 junction transistors.

Meanwhile in America the surface barrier transistor was in vogue, being used in the MIT Lincoln Laboratory's TX-0 computer (first working in 1956) and the Philco Corporation's TRANSAC S-1000 computer (first working in 1957). Both these machines were considerably larger and faster than any British transistor computer, and more was to follow. At about this time news began to come through of two very high-performance American transistor projects known as LARC and STRETCH. Clearly, Britain was falling behind.

By the autumn of 1956 Tom Kilburn and his team at Manchester had begun work on another transistor computer called MUSE – a 'microsecond engine'. This was an ambitious project[28] which aimed at computing speeds approaching 1 microsecond per instruction. At about the same time the Brunt Committee and Lord Halsbury of the National Research Development Corporation were also trying to initiate a British high-speed computer project, spurred by reports of the massive scale of

Fig. 9.2 The Manchester University experimental transistor computer, operational in November 1953 and therefore the world's first transistor stored-program machine to work. The magnetic drum store on the left of the photograph was re-cycled from an earlier computer project (as may be seen by comparison with figure A2.1).

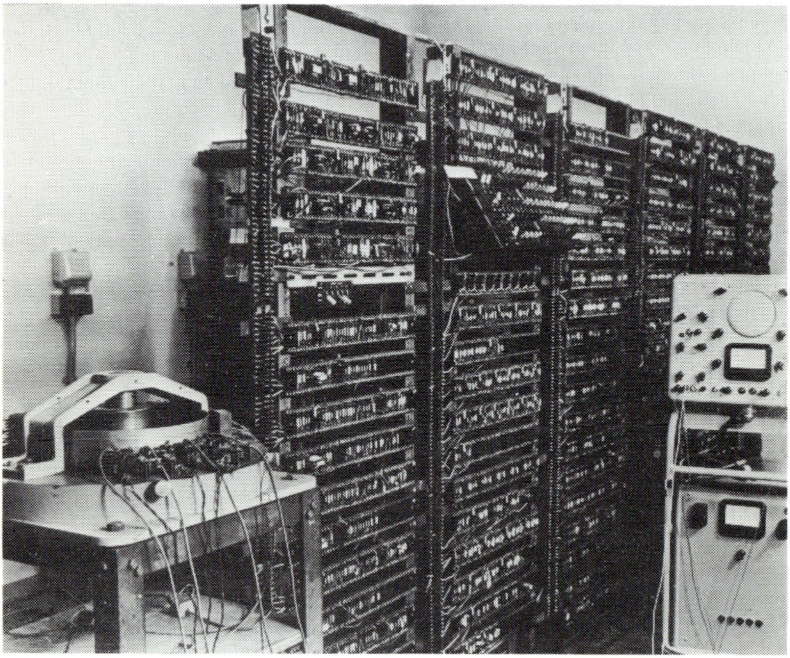

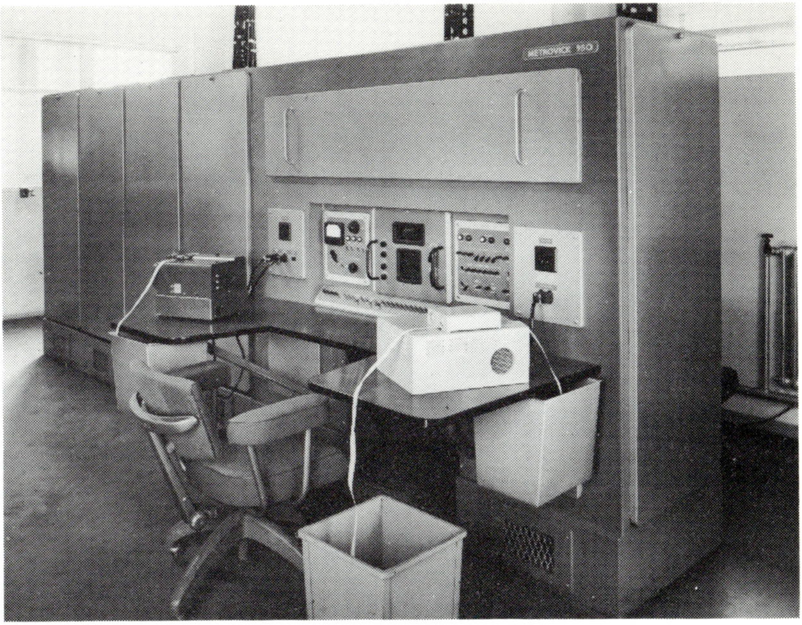

Fig. 9.3 The Metropolitan-Vickers MV950, thought to be the first transistor computer to become commercially available (1956).

the Univac LARC and IBM STRETCH computer developments in America. Like many committees before and since, the NRDC was unable to find agreement on exactly who should design such a computer for Britain. From Manchester University's view, neither government nor industrial support was forthcoming for their own MUSE proposal, and so Kilburn decided to go ahead independently with a limited prototype, using internal resources including a Mark I computer Earnings Fund. Amongst Kilburn's team, D. B. G. Edwards was responsible for the hardware side and R. A. Brooker for software. By good fortune, Ferranti Ltd decided at the end of 1958 to support the project. By 1959 the computer had been re-named ATLAS and was thereafter developed as a joint University/Ferranti venture under Tom Kilburn. Perhaps the main Ferranti design contribution to the project lay in the stalwart work of David Howarth on the Atlas Supervisor – considered by many to be the first recognisable modern operating system.

Fig. 9.4 M. J. Lanigan and Tom Kilburn with a section of the transistorised MUSE computer in 1959. This University prototype was transformed into the Ferranti ATLAS, believed to be the world's most powerful computer at its inauguration by Sir John Cockroft in December 1962. ATLAS introduced several novel concepts into computer design.

ATLAS was not an 'early' computer so a full description is out of place here. Suffice it to say that at its official inauguration on 7 December 1962 it was considered to be the most powerful in the world. Its fastest instructions took 1.59 microseconds and, due to features called virtual storage and paging, each one of many simultaneous users could imagine he had up to one million words of storage space at his disposal. ATLAS pioneered many concepts which are in common use today.

ATLAS was what might be called a 'supercomputer', and consequently had a limited market. At the other end of the spectrum several successful small and medium British transistor computers had meanwhile been developed. Amongst these were the Elliott 803, EMI EMIDEC 1100 and the Ferranti SIRIUS, first delivered in 1958, 1959 and 1960 respectively. There was a long overlap period between valves and transistors, and we should now go back in time a few years to catch up with concurrent valve-based projects.

10
DEFENCE COMPUTERS

The Government's Bletchley Park code-cracking operations have been alluded to earlier. The need to keep ahead in the cryptanalysis stakes must have continued after the war, though details are not available to the general public. Successors to COLOSSUS were built, but it is not thought that these developments contributed in any significant way to the design of general-purpose stored-program computers.

Other fields of defence activity involving digital techniques were the tracking and telemetry problems associated with guided weapons, etc. The principal stored-program computer development here was the MOSAIC (Ministry of Supply Automatic Integrator and Computer) project, which was implemented between 1947 and 1954 by a Post Office team led by Dr A. W. M. Coombes.[29] NPL contributed to the mathematical specification and the All-Power Transformer Co. helped with the manufacture and assembly. Parts of the MOSAIC project are still secret, it having been used for processing radar tracking data in experiments on aircraft. Briefly, the work included building two towable 'data recorders' which produced 3-inch-wide paper tape, together with a very sophisticated display and tracking system. The computer itself had storage for 1024 40-bit words in mercury delay lines, involving nearly a ton of triple-distilled mercury. The physical layout of the delay-line tanks was copied from the scheme used at Cambridge University for the EDSAC. The man who had helped to build COLOSSUS was not afraid of thermionic valves: Coombs used 6000 of them in MOSAIC which, together with 2000 germanium semiconductor diodes, gave a total power dissipation of 60 kilowatts. Although not the earliest of early British computers, MOSAIC was arguably the largest.

The MOSAIC was housed at Malvern and, whilst it was no doubt very valuable in defence circles, its design had little influence on the mainstream of computer development. In view of the price of mercury, perhaps this was just as well.

From 1947 to 1953 Malvern also housed Dr A. M. Uttley's Telecommunications Research Establishment team developing the TREAC computer.[30] TREAC was the only *parallel* computer being designed in Britain

Fig. 10.1 Some central units of the MOSAIC computer, built by the Post Office for a secret Ministry of Supply defence project at Malvern. MOSAIC was working in about 1953 and in terms of electronic components was the largest early British computer.

Fig. 10.2 The Telecommunications Research Establishment's TREAC computer, also operational in 1953. This was the first *parallel* electronic computer in the country. The boxes along the lower right side of the machine contain Williams storage tubes. Note the soldering iron conveniently to hand on the left!

at the time. It had a 512-word Williams tube store backed by a drum, and achieved an addition time of 40 microseconds. Its instruction repertoire was similar to that of the Manchester Mark I, except that there were no index registers. Another difference from the Manchester computer was that the CPU was synchronised to the drum and not vice versa (see Appendix 2). TREAC was notable for the care taken to ensure reliability: exhaustive component tests were carried out during the design phase and the CPU was initially planned with a scheme of 'self-checking' logic.

It should be mentioned that research establishments such as TRE had for some years been developing other 'computers'. These were mainly *analogue* devices for simulating aircraft behaviour, etc. and in the context of our story they have little bearing on the history of stored-program *digital* computers. It is necessary to make the distinction in order to focus attention on the type of machine we now simply call a 'computer'. Analogue devices were, and are, very valuable for certain classes of problem, although it is true to say that their importance is diminishing.

All the early computer projects described so far have been associated with Universities or government research establishments. There was another independent line of development going on in the firm of Elliott Brothers, inspired originally by their work for the Admiralty on anti-aircraft real-time fire control.

11
ELLIOTT BROTHERS

The first British company to become seriously involved with digital computer technology was Elliott Brothers, a London-based firm which had been manufacturing scientific apparatus since 1801. During the war Elliott Brothers had been supplying a great deal of electro-mechanical gunnery control equipment for the Navy. In 1947 Elliott started its Borehamwood research laboratory under J. F. Coales with several naval contracts, including one for the MRS5 advanced digital real-time fire control system. This work involved the construction of a special-purpose computer or on-line digital control system, designated the '152' (Reference 31). The 152 included a fixed-program store based on the principle of the flying-spot scanner and Williams tubes for the working store. The number representation was serial, but parallel arithmetic processing was employed to give a very fast multiplication time of 60 microseconds. Another naval contract involved a special-purpose machine called the 153 which employed nickel magneto-strictive delay lines backed by a fixed-head disc. An interesting feature of the 152 and 153 was the use of miniature thermionic valves and modular circuits mounted on glass printed circuit boards. The logic circuits, developed by C. E. Owen, were to become the foundation for the Elliott 400 series and subsequently the Ferranti PEGASUS computer.

The MRS5 did not involve a truly general-purpose computer. In the end the contract was terminated by the Navy in favour of an analogue system which was being developed concurrently elsewhere. The project did, however, stimulate interest within Elliott Brothers and from the related research came a general-purpose stored-program computer called NICHOLAS. This had a 1024-word nickel delay-line store, an add instruction time of about 10 milliseconds, and first ran a program in

Fig. 11.1 An Aladdin's cave of electronic wizardry: the type 152 naval gunnery control computer, begun at the Borehamwood laboratories of Elliott Brothers in 1947. This special-purpose real-time computer had a fixed program store, and its initial task was to analyse the performance of the radar set via data recorded on photographic film. Five of the six digital cameras for reading film cassettes may be seen in the foreground of the picture. Out of a related Admiralty contract came the basic modular design for several subsequent general-purpose computers.

December 1952. It was used successfully at the Borehamwood Laboratories to carry out trajectory calculations, etc. for a number of years, during which time a very convenient programming system was developed. Technical details of NICHOLAS are given in Appendix 2.

Meanwhile the company's Computing Division under W. S. Elliott received an NRDC contract in September 1950 to study the application of printed circuits and other Borehamwood technology to general-purpose computers. This culminated in a report dated January 1952 which, in April of that year, led to NRDC placing a contract with Elliott Brothers for the construction of a small prototype machine. This became the Elliott 401 computer, for which the detailed design team was led by A. St Johnston. It was completed in April 1953 and exhibited that month at the Physical Society Exhibition in London. The 401 had a 1000-word disc store and nickel delay line central registers. It inherited the modular

Fig. 11.2 The Theory Laboratory at Borehamwood, receiving a contract to compute trajectories, realised that the easiest and most interesting way to solve the tedious calculations was to build themselves a computer. NICHOLAS was the result. It was in parts literally home-made, and the wooden frame and consequent fire extinguisher may be seen in this photograph. NICHOLAS performed valuable service from 1952 to 1958.

plug-in philosophy from the MRS5 project. Elliott Brothers then developed the 402 production version (first delivered in 1955) and other machines in their successful 400 series. The 405 (first delivered 1956) was a general business machine featuring bulk storage on magnetic film and other interesting features relevant to commercial data processing. The prototype 401 was owned by the NRDC, who moved the machine temporarily to Cambridge and continued the development there until March 1954, when it was handed over to the Rothamstead Experimental Station.

In 1953 the 401 team split in two, with W. S. Elliott and others leaving to join Ferranti's London branch. Here they put their modular packaging experience to good use in the design of the Ferranti PEGASUS computer, first delivered in 1955 as described in Chapter 14. The design emphasis for both the 400 series and PEGASUS favoured the use of standard,

Fig. 11.3 The Elliott 401 computer (1953), built under a National Research Development Corporation contract to promote modular circuit techniques. The resulting plug-in packages here give the appearance of books in a book-case. The magnetic disc store and its motor may be seen in the upper right compartment in the photograph. The 401 was the progenitor of a family of successful Elliott and Ferranti modular computers.

interchangeable circuit modules. This obviously made sense for volume production, and contrasted with the 'hand-built' methods used for the larger Ferranti Mark I. Modularity in varying degrees was quickly adopted by all computer manufacturers.

Besides the 405, the team which remained at Elliott Brothers later developed the 800 series of transistorised machines which were in many ways forerunners of the modern minicomputer. The Elliott 802 was first delivered to customers in about 1958. (The American PDP/8, manufactured by the Digital Equipment Corporation, is generally regarded as the first widely used minicomputer proper. In Britain PDP/8s began to arrive in about 1965 and it was not until 1969 that the number of PDP/8 installations overtook the number of Elliott 803s – see reference 50.)

12
PIONEERING SMALL COMPUTERS

With the exception of machines such as the Elliott 402 and the MV950, the computers discussed so far were all considered 'large'. That is to say, the financial outlay and manpower training invested in each computer installation was relatively large for the organisation concerned. From the beginning there were also some intentionally small projects.

A very early pioneer in small computers was A. D. Booth of Birkbeck College, London University. He had spent six months with von Neumann's computer group at Princeton University in 1947 and had returned to England to design ARC, the Automatic Relay Computer. He was supported for a time by the British Rubber Producer's Research Association. ARC was envisaged as a *parallel* stored-program computer, but development of a suitable store caused some problems. Although the arithmetic section of ARC was working in the spring of 1948, Booth's magnetic drum store containing 256 20-bit words was not fully in action until some time later.[32, 33] ARC was eventually equipped with an electro-mechanical store for 50 numbers, together with a pluggable sequence unit for 300 instructions, and operated successfully as a sequence-controlled calculator. (It should be added that Booth was short of funds and was obliged to work more or less as a one-man-team for much of the time, with assistance on the programming side from Miss Kathleen Britten whom he subsequently married.)

By 1949 Booth had moved to thermionic valves, where his emphasis was on the design of low-cost computers which could be, and were, attractive to smaller scientific organisations. After a brief experiment with SEC, a Simple Electronic Computer, he designed the APE(R)C series of All-purpose Electronic (Rayon) Computers. (The 'Rayon' signified the sponsorship of the British Rayon Research Association; other

Fig. 12.1 A. D. Booth, an early pioneer of small computers, seen here with some pattern recognition apparatus at Birkbeck College in 1959. Booth's ingenious ideas transcended the comparatively modest resources at his disposal.

sponsors were indicated by inserting their initials in parenthesis.) The APE(R)C had a re-designed magnetic drum built by S. J. Booth, A.D.'s father, and contained 415 thermionic valves. APE(R)C was operating with limited storage in July 1952, and with a complete store some months later.[34]

The British Tabulating Machine Company became interested in Booth's work and derived their HEC (Hollerith Electronic Computer) from his APE(R)C project. The nomenclature arose because BTM punched-card equipment had always been known as 'Hollerith machines', in recognition of their historic link with Herman Hollerith and to distinguish them from 'Powers machines'. A prototype HEC was exhibited at the Business Efficiency Exhibition in London in 1953. HEC was first marketed as the BTM 1200 computer in 1954, in which year five were sold. The first business data processing version, the BTM 1201, had an enlarged 1024-word drum store and was first delivered in 1956; 70 such machines were eventually sold.

Fig. 12.2 The APE(R)C computer at Birkbeck College, London University, in about 1952. To the left of the photograph is the teleprinter input/output equipment; the right-hand tall 'post office' rack contains a small drum store about two thirds of the way down; the box on the right of the photograph is the power supply. This project was the inspiration for the British Tabulating Machine Company's HEC computer, later marketed as the BTM 1200 series.

Fig. 12.3 The ICCE relay computer at Imperial College, London University, in about 1952. The electro-mechanical relays are contained behind the horizontal rows of covers. ICCE was the largest of the early British relay computers.

Besides Booth's ARC, there were three other relay computer projects of note, all described in appropriate chapters of reference 16. In philosophy, they belonged to the field between calculators with and without the stored-program feature. This in-between area included machines called sequence-controlled calculators which generally held their instructions on some form of semi-permanent external 'storage' such as punched paper tape. Of the three relay computers under discussion, one was at Imperial College, London, under K. D. Tocher and S. Michaelson, another at the Royal Aircraft Establishment (RAE) Farnborough under Dr S. H. Hollingdale and E. J. Petherick, and the third at the Atomic Energy Authority's Harwell establishment under E. H. Cooke-Yarborough and R. C. M. Barnes. The Imperial College machine (called ICCE) employed 20-bit words, with the main storage being external to the main computer, on paper tape. The Farnborough machine was decimal and also envisaged a certain amount of external storage. Originally based on relays, the Farnborough machine eventually went over to a drum store and decatron valves. The Harwell machine was also decimal and initially had 40 8-digit decatron registers for internal storage. Although it could on occasions act as a true stored-program computer, this was not its normal mode of operation. It had a multiplication time of between 5 and 10 seconds, and first worked in April 1951. The Farnborough machine was eventually known as RASCAL (RAE Sequence Calculator) and was still under construction in 1953, by which time the Plessey Company were involved in the development. RASCAL was never actually completed, being overtaken by events at RAE such as the arrival of an English Electric DEUCE. The Imperial College machine was the largest and fastest of the three relay computers, and was working by about 1950. Neither of the three machines seems to have had any direct influence on subsequent industrial developments.

Returning to valves, a small machine with a Dutch pedigree was built by Standard Telephones and Cables Ltd, Monmouth, in the mid-1950s. This was the Stantec ZEBRA, based on a design of W. L. van der Poel from the University of Delpht in Holland. ZEBRA had a main store of 8192 33-bit words on a magnetic drum. When first delivered in about 1957 it cost approximately £23 000, thus being amongst the cheapest general-purpose computers then on sale.[35] A total of about 20 ZEBRAs were installed in Britain.

We have so far been concerned only with digital computers. Throughout the period under discussion *analogue* computers were being built and used, especially for problems related to aircraft design. One of the earliest examples of a *hybrid* computer, that is to say a machine in which a stored-program digital computer is combined with an analogue section, was built between 1953 and 1955 by Smiths Aircraft Instruments Ltd of Cheltenham. This computer was developed by a team under L. Dilger to cope with the large amounts of simulation data encountered in certain problems associated with blind landing systems, etc. The digital section had an accuracy of four decimal digits, using decatrons and a drum for storage. Input/output was via a teleprinter. The project, which was named SECA (Smiths Electronic Calculating Analyser), was intended to complement an analogue computer of advanced design called SEDA (Smiths Electronic Differential Analyser). The development of both SEDA and SECA was undertaken to provide simulator facilities for guided weapons. Unfortunately, certain key contracts were cancelled and

SECA was never fully operational. SEDA, however, continued to be used for about ten years for the development of automatic flight control systems, including those for automatic landing.

Fig. 12.4 A section of the relay computer designed at the Atomic Energy Authority's Harwell establishment in 1951. This machine gave faithful service for many years, being donated to Wolverhampton College of Technology in 1957. It was then re-named WITCH (Wolverhampton Instrument for Teaching Computation from Harwell) and continued to initiate the uninitiated into the mysteries of computing until its retirement in 1973.

13
LEO AND ENGLISH ELECTRIC

The connections between Cambridge University and the J. Lyons catering company, and between NPL and the English Electric Company, have already been mentioned briefly. We should now do more justice to these enterprises and describe how the two companies eventually amalgamated in the early 1960s.

Fig. 13.1 The LEO (Lyons Electronic Office) computer under construction in 1949/50 at the Cadby Hall premises of the J. Lyons catering company in London. Derived from the Cambridge EDSAC, LEO and its associated data handling equipment was developed for the novel purpose of automating office procedures such as payroll processing, at a time when the idea of an 'automatic computer' was regarded by many as a scientific oddity.

Fig. 13.2 The completed LEO computer, showing the central processor bays in the background to the right of the photograph and the operator's console and input/output equipment in the foreground. LEO was finally switched off in January 1965, at which time it was said to be the oldest operating electronic computer. It had given more than 13 years of continuous service.

The pioneering spirit of J. Lyons & Co. cannot be over-emphasised. Although obviously inspired by the research at Cambridge, this essentially non-technical catering company had taken the first formal steps to build its own computer before the Cambridge EDSAC had even been proved to work. Raymond Thompson took the lead at Lyons. The following quotations, taken from his unpublished 'LEO Chronicle', give some highlights in the story of the Lyons Electronic Office (LEO). They tell of the cooperation with Cambridge; of a magnetic tape input/output system and elaborate converter/reconverter system built for Lyons by Standard Telephones and Cables Ltd, and of its eventual abandonment because of unreliability; of its replacement by a system based on card readers and punches provided by the British Tabulating Machine Co.; of the final completion of a computer which contained almost 7000 valves and included 2048 18-bit words of fast storage and two line-printers.[36] Let this unedited selection of entries from the Chronicle speak for itself:

25 February 1947:
Following a suggestion from Standingford that Thompson and he should study the possibilities of electronic calculators in the office during their visit to the USA, a letter is written to Dr Goldstine of Princeton University.

12 May 1947:
Thompson and Standingford visit Princeton.

19 May 1947:
Professor Hartree of the University of Cambridge, on hearing from Goldstine of our enquiries, writes to invite us to meet him and Dr Wilkes.

24 July 1947:
Thompson and Standingford visit Cambridge and see EDSAC for which the main prototype panels are already being tested.

20 October 1947:
A report is submitted to the Board proposing that the Company should take an active part in promoting the commercial development of electronic calculators.

13 November 1947:
The Board decide to make a donation of £2,500 to Cambridge and to lend Lenaerts for six months to Dr Wilkes.

28 September 1948:
A recommendation is made that we should proceed to engage an electronic engineer in spite of the fact that EDSAC is not yet working.

13 December 1948:
J. M. M. Pinkerton is offered the post of electronic engineer.

17 January 1949:
Pinkerton joins the company, first of all studying clerical procedures and then going to Cambridge to study EDSAC for a few weeks.

8 March 1949:
A small workshop is set up in the lodge of St. Mary's College [part of the Lyons Cadby Hall premises in London] to prepare experimental circuits.

8 April 1949:
Report No. 2 is submitted to the Directors indicating the nature of equipment required to feed in data and feed out results at a speed comparable with that of the calculator and suggesting that we should approach a firm such as Standard Telephones to carry out certain development work for us. It is also proposed that we should contact a manufacturer for the wiring of panels for valves for us and that we should get an assistant for Pinkerton and have the use of a draughtsman for the mechanical drawing work.

11 July 1949:
Wayne Kerr Laboratories Ltd sign an agreement to produce panels of electronic circuits to our specification.

31 August 1949:
Report No. 5 is submitted pointing out the need for research into the way in which clerical work must be organised for the calculator and into the technique of programming; as a result a start is made on a systematic analysis of clerical work and the basic techniques of programming.

14 September 1949:
The name LEO (Lyons Electronic Office) is given to the calculating machine – suggested by Mr Simmons.

5 October 1949:
The Coventry Gauge and Tool Company is asked to manufacture the large delay battery tubes.

25 January 1950:
Following discussion on the Standard Telephones report we receive an estimate giving a maximum cost of £16,000 for auxiliary

equipment required for the pilot LEO and a schedule of delivery dates indicating that the delivery of the equipment should be complete by January 1951.

16-31 October 1950:

Pinkerton spends a lot of time at Enfield in an attempt to accelerate production.

15 February 1951:

H. R. H. Princess Elizabeth visits Cadby Hall and sees LEO carrying out a simple test programme.

17 April 1951:

Demonstrations are given to the Directors showing LEO doing clerical work.

5 September 1951:

The Cadby Hall Bakeries Job is done by the calculator, producing accurate results. It is not yet possible to do this job regularly, however, owing to the unreliability of the character generator.

5 October 1951:

In view of the delay in deliveries of equipment from Standard Telephones, Messrs. Simmons, Thompson, Standingford, Pinkerton, Kaye and Caminer discuss the possibility of developing other input and output systems using paper tape; a programme of development is drawn up.

16 November 1951:

Brigadier Hinds writes to offer us £300 for carrying out some ballistic computations for the Ministry of Supply.

10-12 January 1952:

The calculator is given a reliability test producing tax tables for the current year. The test ran over 59 hours and during 51¼ hours of this time useful work was produced giving an operational efficiency of 87%.

14 January 1952:
Contract received for daily ballistic computations.

9 July 1952:
Have a meeting with B.T.M. to discuss the basic requirements of input and output equipment to be attached to LEO.

7 October 1952:
The Management agree that at an additional expense of £10 000 we shall build our own high-speed input and output equipment as an alternative to the Standard Telephones equipment in case this does not prove to be reliable.

1 January 1953:
Successful full scale trial of payroll programme is carried out using both converter and reconverter. There were minor computer faults and flaws in the programming which were later amended.

6 May 1953:
The first tabulator is delivered from B.T.M.

25 August 1953:
Following a further meeting with Mr McVie and Mr Hinton of Standard Telephones on the 19th August, Mr Simmons writes to Mr McVie informing him that we have decided to abandon the project, and asking him to arrange for the equipment to be dismantled and removed.

23 December 1953:
The decoration of the Computer Room is completed. A party is held in the No 1 Dining Room to celebrate the completion of LEO I.

5 November 1954:
Teashops General Report for Wembley teashops has entry: 'The head staff at this shop would like to give thanks for LEO. This is a wonderful time saver, and work saver, and we are grateful for it'.

In November 1954 Lyons founded a new company, Leo Computers Ltd, to build and sell a new computer called LEO II. This had 1024 39-bit words of mercury delay-line store, backed by up to four drums and some magnetic tape decks. LEO II was first delivered in 1957 and about 13 were sold. This machine was followed by various versions of LEO III from 1962 onwards, about 100 of these finally being built.

In 1963 the data processing and control division of English Electric Ltd and Leo Computers Ltd came together as English Electric Leo Computers Ltd. In the next year English Electric bought out Lyons' share, thus ending a fascinating connection between teashops and computers!

The English Electric Company, a well-established manufacturer of electrical machinery and electronic equipment, first became interested in digital computers through contact with the Pilot ACE group at NPL.[37] Sir George Nelson (later Lord Nelson), the Chief Executive of English Electric, was at that time a member of the NPL Executive Council. He and his company's Research Director (J. K. Brown) quickly realised the implication of the NPL proposal to develop a computing engine. This contact was formalised in January 1949 when A. C. D. Haley was recruited by English Electric to head a team of about a dozen engineers and technicians on loan to NPL. The English Electric team cooperated with the NPL group under J. H. Wilkinson and E. A. Newman to produce the Pilot ACE which went into full-time service in 1952, as described in Chapter 8.

Computer design was continued by the company at its Nelson Research Laboratories in Stafford, and later at a special factory a short distance away at Kidsgrove. This activity was initially inspired by the company's own need for in-house digital computing facilities, but early enquiries from outside organisations such as the Royal Aircraft Establishment quickly put the project on a commercial footing. The first machine to be marketed was the English Electric DEUCE, first delivered in the spring of 1955. Like the Pilot ACE, DEUCE[38] used mercury delay lines backed by a magnetic drum for storage, punched cards for input, and operated at the comparatively fast digit rate of 1 megacycle. Its internal structure was the same as the Pilot ACE. However, the physical construction was considerably developed so as to give good reliability

Fig. 13.3 The control desk of an English Electric DEUCE computer. First delivered in 1955, this machine was the production version of the NPL Pilot ACE. Input was via the punched-card reader to the left of the photograph. The black hand-punch lying on the right of the control desk was no doubt only for quick card edits.

and easy maintenance. DEUCE contained 1450 valves, thus being nearly twice the size of the NPL prototype. From late 1955 onwards English Electric began selling an improved version known as DEUCE 2, followed in 1957 by a DEUCE 2A; these featured amongst other things a re-engineered input/output system. The company sold about 31 DEUCE 1 and 2 machines between 1955 and 1964.

During the late 1950s English Electric developed three smaller computers, arising initially out of the company's involvement with digital data logging equipment for steel rolling mills, nuclear power generating plants, etc. The first of these computers, known as the KDN2, used transistor circuits and was available from 1962. The design engineer most closely associated with this project was K. Chisholm. From the KDN2 two further small computers were developed: the KDF7 for general process control applications and the KDF6 for commercial data processing. These were all 16-bit word length machines, and approximately 30 of them were sold.

Concurrent with this activity, English Electric embarked on two major computer projects. Firstly, as a result of a long-standing technical arrangement with the RCA company of America, English Electric built a version of the RCA501 computer which was known as the KDP10. This was a variable word-length machine intended for commercial data processing applications. The KDP10 was first delivered in 1961, after which it was upgraded and re-designated the KDF8; about ten machines were sold.

The second, and historically more interesting large computer to emerge from Kidsgrove was the KDF9. This high-speed transistor machine[39] was developed by a team under A. C. D. Haley, of which the leading light was R. H. Allmark. The KDF9 is remarkable because it is believed to have been the first zero-address instruction format computer to have been announced (in 1960[40]). It was first delivered at about the same time (early 1963) as the other famous zero-address computer, the Burroughs B5000 in America. Like many modern pocket calculators, a zero-address machine allows the use of Reverse Polish arithmetic; this offers certain advantages to compiler writers. It is believed that the attention of the English Electric team was first drawn to the zero-address concept through contact with GEORGE (General Order Generator[58]), an autocode programming system written for a DEUCE computer by the University of Sydney, Australia, in the latter half of the 1950s. GEORGE used Reverse Polish, and the KDF9 team were attracted to this convention for the pragmatic reason of wishing to enhance performance by minimising accesses to main store. This may be contrasted with the more 'theoretical' line taken independently by Burroughs.[60] Besides a hardware nesting store or *stack* – the basic mechanism of a zero-address computer – the KDF9 had other groups of central registers for improving performance which gave it an interesting internal structure. Certain aspects of the stacking idea are now to be seen in many modern computers. Although not pioneering in the strictly chronological sense of machines such as EDSAC, the KDF9 is nevertheless worth remembering as a very stimulating example of innovative design. About 29

KDF9 computers were sold between 1963 and 1969.

In 1964 English Electric Leo Computers Ltd acquired the Marconi Company's commercial and scientific computer interests. The enlarged company was accordingly re-named English Electric Leo Marconi Computers Ltd. In 1967 there was a merger with Elliott Automation and, no doubt hesitating to add yet another word to the company's title, the name was simplified to English Electric Computers. At about this time the firm sold its first System 4 machine. The System 4 was a range of computers whose structure was compatible with that of the IBM System/360 series, and in fact based on RCA's SPECTRA series. In 1968 ICT and English Electric Computers merged to form ICL, as described in the next chapter.

14
FERRANTI LTD, ICT AND ICL

Ferranti Ltd have been pioneers in electrical equipment since the days of the founder, Sebastian de Ferranti (1864–1930). An initial appraisal of the potentials of stored-program computers was undertaken by the company in 1948 and this conveniently prepared the ground for the government-sponsored link with Manchester University, described in Chapter 7. With the impetus of government money and the access to University research, Ferranti Ltd quickly became the largest British stored-program computer manufacturer. When they started selling computers in 1951 there was no real competition; the only impediment was the inertia of the customers.

Table 1 *Ferranti valve Machines of the 1950s*

Computer	*Date first delivered*	*Total number sold*
Mark I	1951	2 (1)
Mark I Star	1953	7 (2)
PEGASUS 1	1956	26 (3)
PEGASUS 2	1959	12 (1)
MERCURY	1957	19 (6)
PERSEUS	1959	2 (2)

(The figures in parenthesis show the number of machines exported)

The Ferranti valve machines which appeared on the market during the 1950s are shown in Table 1.

The Ferranti PEGASUS[41] deserves further mention. This was a highly

Fig. 14.1 Degrees of modularity: plug-in units from an English Electric DEUCE (left) and a Ferranti PEGASUS computer, both machines designed to facilitate quantity production and allow rapid maintenance.

successful offshoot from the packaging ideas of the Elliott 401 computer referred to in Chapter 11. In 1953 and 1954 Christopher Strachey had worked for the National Research Development Corporation on redesigning the 401 and as a result had persuaded NRDC that a more ambitious machine could be developed with Ferranti.[42] Amongst the design objectives recommended by Strachey were: (*a*) optimum programming was to be avoided 'because it tended to become a time-wasting intellectual hobby of programmers'; (*b*) the needs of the programmer were to be a governing factor in selecting the order code; (*c*) the computer was to be cheap and reliable. By the autumn of 1954 a Ferranti design team under W. S. Elliott had set to work in London to build what was at first called the Ferranti Packaged Computer (FPC1), with an NRDC contract for an initial ten machines. (There was some rivalry, to put no stronger word on it, between the newer London and older Manchester sides of the Ferranti computer design effort.)

It should be said that not everyone in the computing fraternity supported the 'package' philosophy. It was understandably criticised as being wasteful of hardware, prone to connector-problems, and requiring a lot of adjustment to account for small differences in circuit parameters between packages. By conservative design of PEGASUS, Elliott avoided the last criticism. Hardware redundancy was said to be only about 10 per cent, and the quick repair-time obtained by having replacement modules amply justified the package approach. The degree of structural modularity also facilitated subsequent improvements to the computer.

PEGASUS had a small 56-word fast store of nickel delay lines, backed by a 5120-word magnetic drum store. There was in addition a set of special registers associated with input/output equipment, etc. which could also be employed to hold frequently used program constants. The word length from the programmer's view was 39 bits and two 19-bit instructions were packed to a word. The spare bit could be used as a break-point marker, to facilitate manual debugging of programs. There was also an extra parity-bit for each word – an early example of this

Fig. 14.2 A highly-polished Ferranti PEGASUS computer on test at the factory prior to delivery in about 1956. Some magnetic tape units may be seen at the left of the photograph, behind the two square paper-tape readers sitting on the table. The PEGASUS internal register structure was the model for many subsequent machines, for example the ICL 1900 series and the IBM System 360.

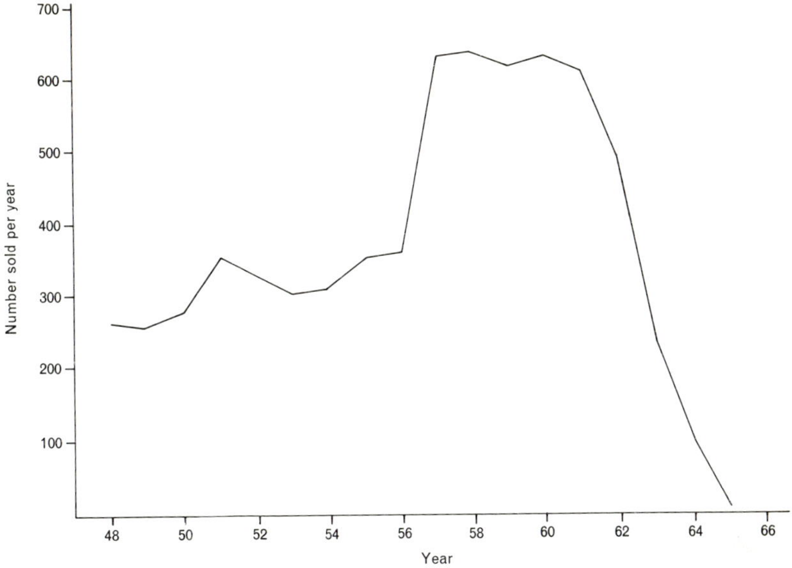

Fig. 14.3 Annual sales of all punched card equipment manufactured by the British Tabulating Machine Company, from 1948–1965. Sales were initially unaffected by the advent of the stored-program computer, especially since most early computers used paper tape for input. In about 1957 computers using cards began to make their presence felt.

form of checking. PEGASUS had a large repertoire of instructions, which often led to relatively compact programs. There were eight accumulators, seven of which could also be used as index registers.

The basic structure of the resulting PEGASUS order code was to have a strong influence on the Ferranti ARGUS, ORION and FP6000 computers and hence on the ICL 1900 series. Strachey had done a lot of his early programming on the Ferranti Mark I. It is interesting to speculate whether his combining of the Mark I's A and B lines into a single register set on PEGASUS was motivated by programming convenience or economics or both. Looking back, we can see a distinction between machines such as the Mark I, Atlas and CDC6600 in which the address-generation and counting registers are kept separate from the main computational accumulator(s), and machines such as PEGASUS, the ICL 1900 and the IBM System/370 in which no such separation occurs. There are arguments in favour of each approach.

Compared with the earlier Elliott 402 computer, the Ferranti PEGASUS operated at about the same basic speed but had a larger store, longer word and considerably more convenient order code. It also cost about twice as much. The first PEGASUS was completed in 1956 and, incidentally, remained in use for the next 13 years. In March 1956 Ferranti used a PEGASUS to initiate the first computer bureau service in Britain, at Portland Place in London. About 38 PEGASUS 1 and PEGASUS 2 computers were sold, making it the most popular of the Ferranti valve machines. It was a common comment on PEGASUS that users 'were not only able to do a lot of work on it; they also felt a strong affection for the machine'.[42]

Concurrent with this activity Ferranti Ltd was designing small computers for defence work. The ARGUS series was one of the outcomes

Fig. 14.4 Annual sales of three classes of computer, for the years covering the merger of the main British computer companies. The graphs typify the following categories: early large and medium scientific machines (e.g. Ferranti excluding Argus, etc. - - -); early small and medium general-application machines (e.g. BTM/ICT 1200 etc. . . .); the evolution of the first two categories into a modern range of business and scientific machines (e.g. ICT/ICL 1900 series——).

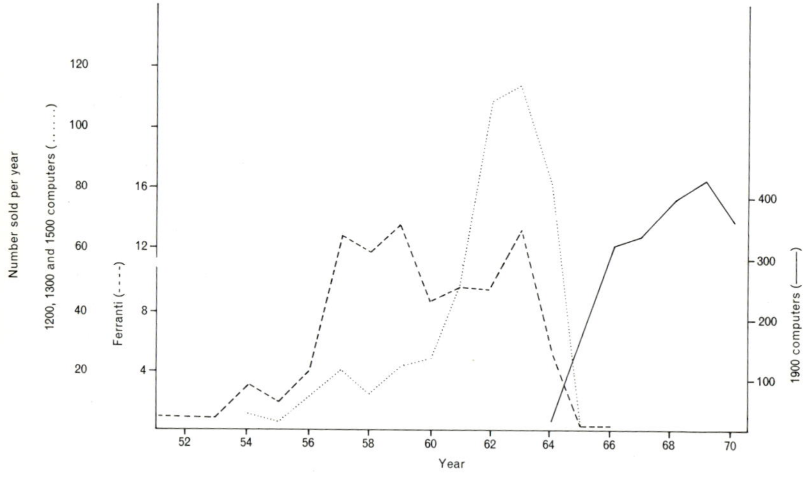

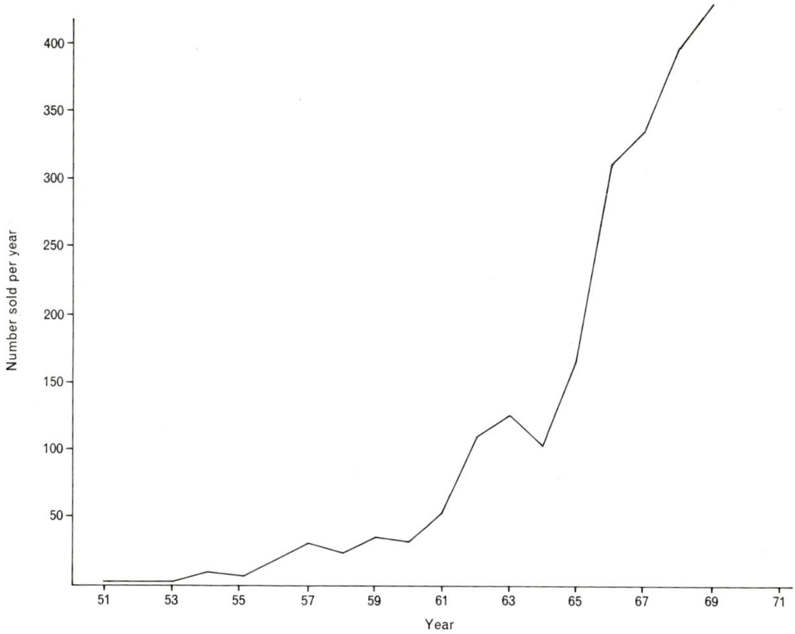

Fig. 14.5 The growth of the computerised society. The three graphs in Figure 14.4 are combined to show that, relatively speaking, sales of Ferranti, BTM and ICT stored-program computers did not really take off until the mid-1960s in Britain.

of this and from 1963 onwards ARGUS was applied to non-military projects in great numbers. In the early 1960s other Ferranti defence computers such as POSEIDON, HERMES and the F1600 were built. Derivatives of the F1600 have been applied to several hundred civil as well as military projects world-wide.

In the late 1950s Ferranti Ltd embarked upon two substantial computer projects. The first was the giant ATLAS, the joint venture with Manchester University described in Chapter 9. This project was full of financial uncertainties, as may be judged from the various estimates of the works cost of the first production ATLAS. These estimates ranged from £375 000 in February 1959, to £650 000 in November 1960, to £930 000 in October 1962. At the same time Ferranti was developing another medium-to-large computer named ORION, which ran into some technical difficulties. These included problems with interconnecting the special circuits which had been designed to economise on the total number of transistors used. By 1963 deliveries of ATLAS and ORION computers had begun, but sales were slow in coming.

With the drain on resources occasioned by these two large computer developments and the growing internecine competition from rival British companies in the mid-range market, some rationalisation of the industry had to take place. Ferranti's main computer interests were sold to International Computers and Tabulators (ICT) in September 1963.

ICT was formed by the merger in 1959 of the two main British punched-card equipment manufacturers: The British Tabulating Machine Company (BTM) founded in 1907 out of the same American parentage which was to produce IBM; and Powers–Samas Accounting Machines Ltd which had been founded (under another name) in 1915 out of the same American parentage which was to produce the Univac Corporation. In 1961 ICT acquired the computer interests of the General Electric Co. Ltd (GEC), and those of Electrical and Musical Industries Ltd (EMI) in 1962. By January 1963 the value of computing equipment installed in the United Kingdom was estimated to be distributed amongst the manufacturers as follows:[42]

Ferranti Ltd	25%
ICT	25%
English Electric Co.	13%
Elliott Brothers	12%
National Cash Register Co. (NCR)	11%
Leo Computers Ltd	7%
Others	7%

English Electric absorbed Leo Computers in 1963, and the Marconi and Elliott computer interests in 1964 and 1967 respectively. ICT and English Electric themselves came together in 1968 to form International Computers Ltd (ICL), thus bringing all the major British computer manufacturers under one roof.

Historically, ICL contained a large component of punched-card equipment manufacturing. It is interesting to see how sales of this type of equipment took a long time to feel the impact of stored-program digital computers. Figure 14.3 shows the sales figures for the punched-card tabulators of BTM from the year 1948 onwards. It is not until 1962 that sales begin to fall, by which time smaller computers such as the ICT 1300 and Elliott 800 series had really penetrated the lower end of the market. Another interesting observation is the dramatic leap in sales, by nearly 100 per cent in 1957.

Figure 14.4 shows the relative growth of stored-program computer installations. Sales of three representative groups of machine are plotted: firstly, all the main Ferranti computers including ORION, SIRIUS and ATLAS, but excluding ARGUS, etc.; secondly, the 1200, 1300 and 1500 series machines produced under the BTM and ICT banners; thirdly, the ICT/ICL 1900 series of computers. The first group represents medium and large computers, with the applications up to 1959 being exclusively scientific and technical.[42] The second group typifies small and medium computers, with sales starting off in the technical area but moving towards commercial and business data processing by the beginning of the 1960s. The third group exemplifies the present-day trend of marketing a compatible *range* of computers, which can embrace both small and large installations and both scientific and commercial applications. (It was no accident that the take-off for ICT 1900 series sales was designed to correspond with the fall in the other two sales curves.)

Several interesting observations may be made on the graphs. Firstly, there was a general surge of investment in data-processing equipment in 1957, but as far as *stored-program* computers were concerned this only really corresponds to an awakening of awareness in the scientific and technical sectors. The wider use of computers did not come about until the early 1960s. This is supported by a further graph (Figure 14.5), in which sales of all the Ferranti, BTM and ICT computers are combined. It is clearly seen that the 'computerised society' did not get under way in Britain until about 1966. It may or may not have been significant that the government founded the National Computing Centre in 1965.

A word should be said about some of the computers introduced in the foregoing graphs. The BTM 1200 series had an interesting pedigree dating back to the pioneering work of Dr A. D. Booth at Birkbeck

College, London University, as described in Chapter 12. The BTM 1300 series arose from original designs carried out by the General Electric Company. The BTM 1300 and ICT 1500 computers were transistorised. The ICT 1500 was based on the American RCA 301 computer. The Ferranti SIRIUS computer, first delivered in 1960, grew out of a machine called NEWT which was the testbed for the special circuits used in ORION. The concept of the ICT 1900 series originated in a paper design sometimes called HARRIAC, drawn up in the autumn of 1961. This was named after Harry Johnson of Ferranti Ltd who chaired a study group of engineers and sales staff which was asked to produce a specification for the successor to the Ferranti PEGASUS. HARRIAC was later dropped in favour of the ORION 2 design. The specification of HARRIAC was passed to Ferranti–Packard of Canada, where it became the basis of their FP6000 computer. The ICT 1900 series in turn grew out of the FP6000 machine.

The 1900 series had a long and successful life, extending almost into the 1980s. By 1968 over 1000 machines had been installed. In October 1974 ICL announced its new range, the 2900 series. This also had an interesting pedigree. Tom Kilburn's team at Manchester University produced their fifth large computer, MU5, between 1969 and 1974. The architecture of the 2900 series owes much to, and has a great deal in common with, MU5.

The history of British computers as recounted thus far has tended towards an engineer's view of the machines – in a word, emphasising their *hardware*. Today the *software* (or set of programs provided by the manufacturer) is normally of more interest to the user than the electronics. In 1950 this was not the case; the mathematically inclined user, if he were to succeed, had to adopt some of the pragmatic optimism of the engineer when faced with the challenge of an early computer and its embryonic software. The next chapter attempts to give the flavour of those early programming days.

15
PROGRAMMING AN EARLY COMPUTER

The user of a modern computer would find life very different on an early machine. Amongst the difficulties common to all vintage computers were: (*a*) *unreliability* and the problem of deciding whether an unexpected result was due to a programming error or an intermittent hardware fault; (*b*) the need to *scale* variables, there being no arithmetic overflow detection, no floating-point hardware, and usually no facilities such as a carry-register to help with multi-length arithmetic; (*c*) *restrictions* due to limited storage and small repertoire of instructions; (*d*) lack of conventional *software support* such as editors and file managers.

Generally speaking, an early programmer rose to the challenge, even becoming quite affectionate towards the computer whose idiosyncrasies he had learnt to master. In the light of hindsight it is now possible to compare the programming idiosyncrasies of various early machines (see, e.g., reference 43). Such comparisons are in no way criticisms of the pioneering designers, who had unimaginable problems and little experience to build on. However, we may ask what, apart from speed and size of storage available, made some early computers seem easier to program than others? Three things stand out: (*a*) the structure of an order code and its repertoire of operations; (*b*) the quantity and quality of standard software such as loaders and subroutine libraries; (*c*) the form of presentation of a program to the machine, which is obviously related to the first two points.

We may get some idea of early programming systems by considering a simple example. Suppose that during the course of solving a problem it was necessary to add two quantities together so as to produce a third. Algebraically this might be expressed as:

$$J = K + L$$

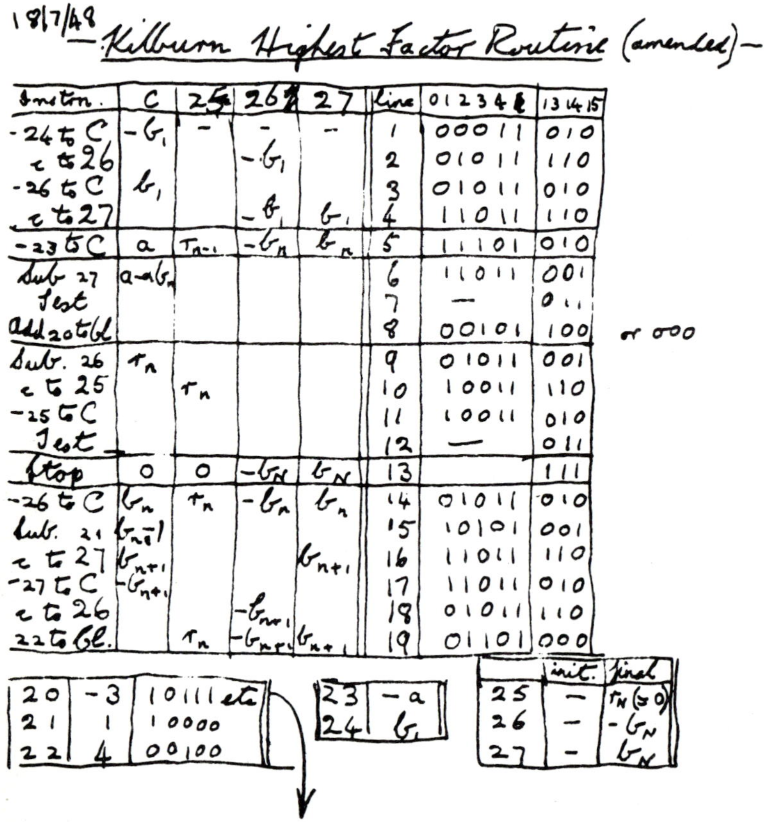

18/7/48 — Kilburn Highest Factor Routine (amended) —

Instrn.	C	25	26	27	line	01234	13 14 15	
-24 to C	$-b_1$	–	–	–	1	00011	010	
c to 26			$-b_1$		2	01011	110	
-26 to C	b_1				3	01011	010	
c to 27			$-b_1$	b_1	4	11011	110	
-23 to C	a	r_{n-1}	$-b_n$	b_n	5	11101	010	
Sub 27	$a-nb_n$				6	11011	001	
Test					7	—	011	
Add 20 to Cl					8	00101	100	or 000
Sub. 26	r_n				9	01011	001	
c to 25		r_n			10	10011	110	
-25 to C					11	10011	010	
Test					12	—	011	
Stop	0	0	$-b_N$	b_N	13		111	
-26 to C	b_n	r_n	$-b_n$	b_n	14	01011	010	
Sub. 21	$b_n - 1$				15	10101	001	
c to 27	b_{n+1}			b_{n+1}	16	11011	110	
-27 to C	$-b_{n+1}$				17	11011	010	
c to 26			$-b_{n+1}$		18	01011	110	
22 to Cl.		r_n	$-b_{n+1}$	b_{n+1}	19	01101	000	

20	-3	10111 etc (or 10100)
21	1	10000
22	4	00100

23	-a
24	b_1

	init.	final
25	—	r_N (= 0)
26	—	$-b_N$
27	—	b_N

Fig. 15.1 A revised version of the world's first working program (see reference 9). This found the highest proper factor of the number placed in store line 23, using a process of repetitive subtraction to achieve the effect of division. The use of relative indirect and absolute indirect jumps (at lines 8 and 19 respectively) is interesting. On conclusion, the program halted at line 13 with the answer displayed at line 27. Folklore has it that this was the one and only program ever written by its originator: little wonder, since it had to be laboriously keyed into the computer in binary!

Modern problem-orientated high-level languages such as Fortran allow the user to write just such a statement in his program. Fortran, however, was not introduced until late 1956. The early computers had to be programmed in so-called machine code, the form of which naturally depended on the computer's structure. For a one-address format computer the statement $J=K+L$ might be achieved by three separate machine-code orders having the following actions:

(1) load K into the accumulator
(2) add L to this
(3) store the result at J.

Fig. 15.2 Early data preparation equipment: an operator punches out a program on 5-track paper tape, prior to it being fed into a computer.

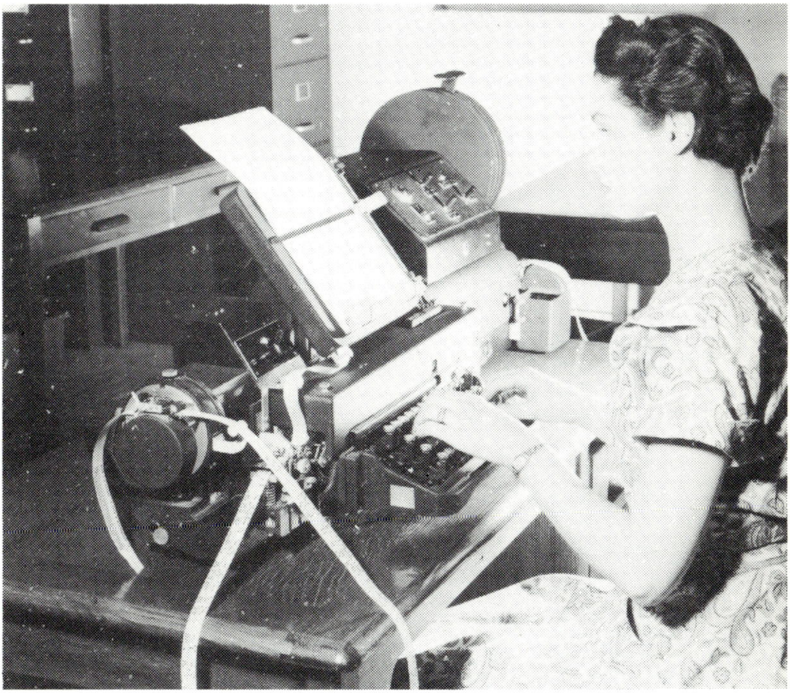

Of course the programmer had to allocate numerical storage addresses to the variables *K*, *L*, *J*. If these addresses happened to be 24, 73 and 66 respectively, then a symbolic representation of the three machine-code instructions might be written:

```
LDA 24
ADD 73
STO 66
```

The symbolic assembly language for a modern one-address format microcomputer may allow the user to write machine-code instructions in just this form. An early computer (which might have had rather less computing 'power' than a modern micro) would not usually have had such a convenient assembly language. The programmer was therefore obliged to present his instructions to the computer in a more primitive form. For example the Cambridge EDSAC instructions to carry out the $J=K+L$ operation could have been written as follows:

```
T1  S  (Transfer acc. to line 1, and clear acc.)
A24 S  (Add into acc. the contents of line 24)
A73 S  (Add into acc. the contents of line 73)
U66 S  (Unload acc. to line 66)
```

(The 'S' indicates that each instruction refers to 'short' (17-bit) EDSAC numbers; the comments in parenthesis are not part of the program. An alternative three-instruction sequence could have been written if the accumulator was initially assumed to have contained zero.) In contrast the 1949 Manchester Mark I had a much more primitive 'assembler' that in effect obliged the user to write his programs directly in binary. Each 5-bit group of a 20-bit instruction was punched or written as the teletype symbol which would give the corresponding bit-pattern. An early Manchester Mark I program thus appeared to the uninitiated as a string of seemingly meaningless teleprinter characters which were much harder to 'read' than the EDSAC programs. The '$J=K+L$' statement might be written as follows for the Mark I:

O//H	(load acc. from line 24)
D@/M	(add into acc. the contents of line 73)
@@/D	(store acc. to line 66)

(As a hint for anyone wishing to decode these, Mark I instructions were in 'backwards binary' with the address to the left.)

Fig. 15.3 The no-man's-land between programmers and maintenance engineers: a page from the users' log-book of a Ferranti Mark I for January 1952 which illustrates the frustration of trying to pin down an intermittent computer fault. The transcription (with explanation) reads as follows: Wed. Jan. 16th 8.30 p.m. Sideways adder [i.e. 'population count' instruction] MSD [most significant digit] not working properly – gives wrong result. Needs to be operated twice to give any result at all – not always – operation quite different according to 'continuous prepulses', 'semi-continuous', 'single' or manual operation. [later] MSD appears to work in programme. 9 p.m.: write power [to the drum] went off though overriding switch and light were on. Cured by switching off overriding switch. OK when switched on again. Thursday Jan. 17th, 17.20 hours. 'No prepulse' phenomena referred to on previous page: definitely not finger trouble this time!!! Witnessed by A. E. Glennie and F. C. Williams!!! Machine unserviceable (from 17.20–17.23?) [Signed] R. A. Brooker. See fault log. [Signed] N.E.H. [the duty maintenance engineer]. 1600–2200 hours: machine behaving very well. [Signed] R.A.B. and A.E.G.

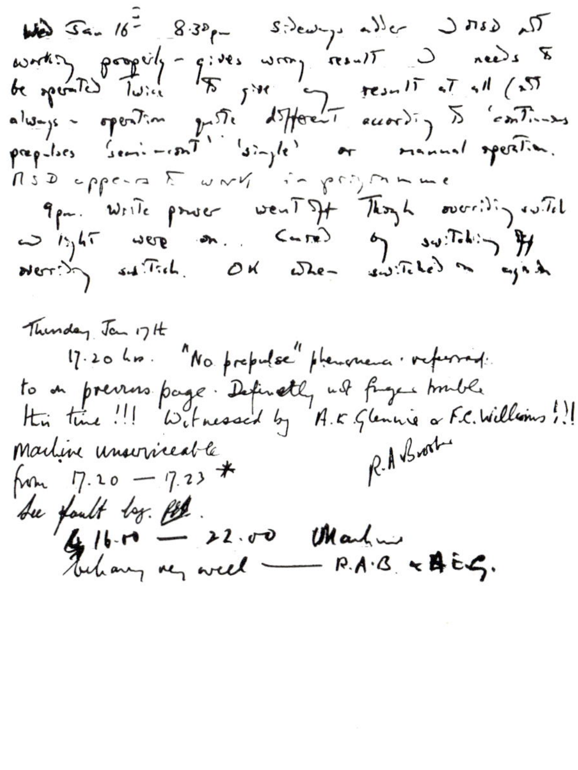

An example of an intermediate programming notation is provided by BINAC, an early American computer described in the next chapter. For BINAC the user had to present his program as a string of octal (i.e. base eight) numbers.[44] Assuming that the instructions 'store-and-clear', 'add' and 'unload' had the octal equivalent 04, 05 and 02 respectively, the BINAC program to perform '$J=K+L$' would have been written as follows:

04001 (Transfer acc. to line 1, and clear acc.)
05030 (Add into acc. the contents of line 24)
05111 (Add into acc. the contents of line 73)
02102 (Unload acc. to line 66)

The programmer would have typed these octal digits at BINAC's input keyboard.

Though they differ in degree of convenience, the above three program conventions are readily understood by the competent modern programmer. There were other early computers whose instruction formats now seem rather curious. A good example is the NPL Pilot ACE and its descendant DEUCE, for which each 32-bit instruction required the programmer to specify seven quantities, namely:

Next instruction source, N
Source of operand, S
Destination of operand and implicit operation required, D
Characteristic (or 'transfer mode'), C
Wait number, W
Timing number, T
Go digit, G

The N field had 3 bits and specified a delay line (containing 32 words). S and D specified a source and destination between which a number of words were to be transferred – (one-word transfers if C=1). The word-address within a particular delay line was obtained as a word-count

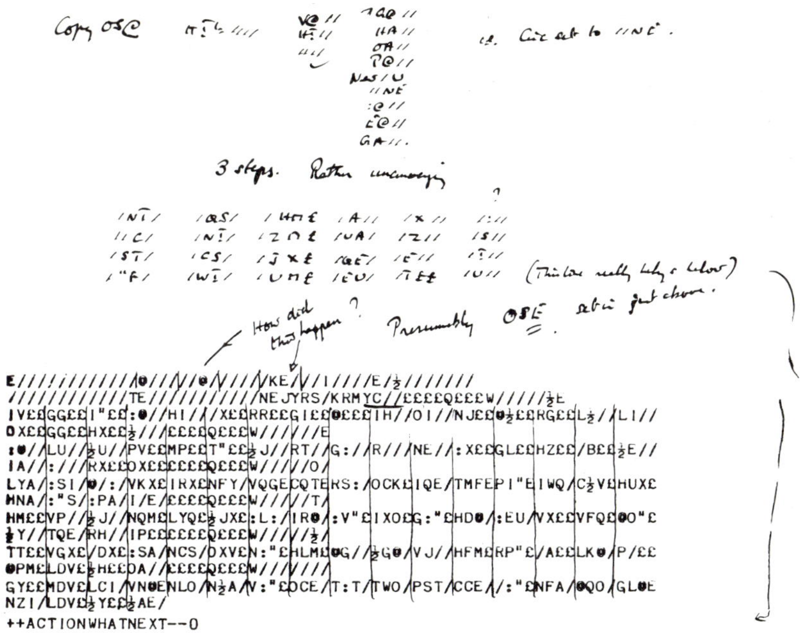

Fig. 15.4 'How did this happen?' The debugging of part of Alan Turing's morphogenesis programs for a Ferranti Mark I was not for the faint-hearted! The photograph shows a fragment of computer print-out with Turing's notes; dated 24 May (1953?)

relative to the position of the current instruction, such word-counts being expressed as a number of 32-microsecond 'minor cycles' by the 5-bit W and T fields of the instruction. (The W and T numbers were actually less by two than the difference in relative position of the words concerned, counting modulo 32.) For a full explanation see for example the paper by Wilkinson.[23] Limiting ourselves to a simplified view of the '$J=K+L$' problem, Pilot ACE would require three instructions of the form:

N	S	D	C	W	T	G	
1	1	16	1	22	31	1	(load acc. from address 24)
1	3	17	1	6	31	1	(add into acc. the contents of 73)
1	16	3	1	30	31	1	(store acc. to address 66)

These three instructions (less the top line and comments) would have been prepared for the Pilot ACE on cards punched in *binary* at one instruction per row, so that a single punched card could hold up to twelve instructions. An ingenious pre-punched four-instruction loader made the input of program cards a simple matter.

The above sequence assumes for simplicity that the three instructions are held in positions 0, 1 and 2 of delay line 1 (hence the 31 in each T-field). It will be noticed that this positioning results in relatively long wait-times in the W-field of each instruction, and such an inefficient program would never have actually been written for the Pilot ACE. So-called optimum programming techniques would have been used as a matter of course, to reduce the wait-times by repositioning some of the instructions. The presence of six short delay lines for frequently used variables could also have helped to avoid long access-times. To modern eyes the Pilot ACE user might ultimately appear to be *microprogramming* the computer, and indeed he had to be aware of the detailed interconnections of the machine. It should however be remembered that the highly original programming strategy for ACE was born in 1945, before most other 'early' computers were even a twinkle in their designer's eye!

To return to more readily understandable computers, it is interesting to compare the order codes of machines which, on the surface, have a similar structure. The BINAC, EDSAC and Manchester Mark I (1949 version) were all one-address format computers, of similar word-length, all designed within a few months of each other, and each being claimed on various occasions as the first stored-program machine to have come into operation. Table 2 gives the modern description of some specimen instructions, and an indication of whether their equivalent was available on each of the three early computers. (The table contains some approximations, which the interested reader is invited to investigate via the original papers.[19, 44, 45])

It will be seen from the table that the Manchester Mark I differed from the other two machines in several respects. It had limited conditional testing orders and shifting orders, but more logic operations and more facilities to aid multi-length working. (The incentive for providing multi-length facilities at Manchester is given in 46, which also describes what is believed to be 'the first routine for a problem with some intrinsic interest to have been run on a "general-purpose" computer'.) Furthermore, the Mark I had two index registers, and a relative jump and indirect jump instead of the absolute jumps of BINAC and EDSAC. To a modern programmer the inclusion of these last three facilities would eliminate the need for self-modifying code, facilitate the generation of position-independent code, and ease subroutine entry and return. On machines such as the Mark I with two levels of storage, the consequent ability to achieve dynamic relocation could obviously be used to advantage. Having said this, it should be stated that phrases such as 'dynamic relocation' and 'position-independent code' were not part of the vocabulary of the first programmers, who were necessarily less ambitious in their use of a computer. The first question, especially with the prototype Mark I, was usually 'will the machine work for long enough for me to get useful results?' This being so, Table 2 is perhaps of more use in tracing the spread of ideas rather than in assessing ease of programming of the machines themselves.

Luckily for the unskilled user, libraries of subroutines were built up for most early computers, so that common operations such as taking a square root or inverting a matrix, could be performed by calling upon a standard library subroutine which an expert had previously written and tested. In the late 1950s high-level languages began to appear, making life even easier for the unskilled user. In the Manchester Mark I Autocode, one of the earliest such languages (March 1954[47]), our '$J=K+L$' statement would have been written simply as:

$$\mathrm{n1=n2+n3}$$

The work of John von Neumann and the EDVAC group in America was the source of several ideas on stored-program computers and it is now appropriate to catch up on transatlantic developments.

Table 2 *Availability of instructions on three early computers*

Type of instruction	*BINAC*	*EDSAC*	*Mark I*
Load accumulator			√
Load accumulator negatively			√
Store accumulator	√	√	√
Store and clear accumulator	√	√	
Clear accumulator			√
Add into accumulator	√	√	√ c
Subtract from accumulator	√	√	√
Load multiplier register		√	√ c
Load multiplicand register and multiply			√ cd
Multiply	√		
Multiply and add into accumulator		√	
Multiply and subtract from accumulator		√	
Divide	√		
Logical AND		√ b	√
Logical OR			√
Logical non-equivalence			√
Shift right by *n*	√ a	√	
Shift left by *n*	√ a	√	√ a
Load index register			√ e
Store index register			
Jump (absolute) unconditionally	√		
Jump (absolute) if accumulator ⩾0		√	
Jump (absolute) if accumulator <0	√	√	
Skip next instruction if accumulator <0			√
Jump (relative) unconditionally			√
Jump (indirect) unconditionally			√
Input a character		√	√ f
Output a character		√	√ f
Stop	√	√	√
Other miscellaneous orders	4g	3h	5j

a BINAC and Mark I shifted by one place only.

b EDSAC 'anded' the contents of the multiplier register with an operand from store, leaving the result in the the accumulator.

c The Mark I had both signed and unsigned versions of these orders; for the signed cases a copy of the operand's sign digit was usually extended into the top half of the double-length (80 bit) accumulator.

d The result of a Mark I multiplication ended up in the double-length accumulator.

e The Mark I had two B-lines (index registers), the contents of B0 normally being zero so as to give the option of unmodified as well as modified addresses for all instructions.

f The October 1949 Mark I allowed programmed transfers to and from I/O devices and magnetic drum by means of a suitably formatted control word.

g The extra BINAC orders consisted of two for manipulating the L register used in multiplication and division, a null op. order and a 'stop if break-point switch set'.

h The extra EDSAC orders consisted of two for rounding off the accumulator (to 16 bits or 34 bits) and one for testing the interface to the printer.

j The extra Mark I orders consisted of five instructions for manipulating the most significant half of the double-length accumulator.

16
MEANWHILE IN AMERICA . . .

In America the actual *implementation* of stored-program computers got off to a slower start, after the early enthusiasm evident in the EDVAC and Princeton draft proposals.[10] This was possibly due to three reasons. Firstly, there existed between 1945 and 1949 a substantial investment in the USA in large-scale electro-mechanical and electronic calculating machines, which were able to satisfy many of the country's immediate computational needs. Amongst these machines were the Bell Telephone Laboratories' series of relay calculators, the Harvard Mark II and Mark III electro-mechanical machines (the latter with eight magnetic drums), IBM's Card-Programmed Electronic Calculators and their huge Selective-Sequence Electronic Calculator (containing 23 000 relays and 13 000 thermionic valves), and of course the enormous ENIAC.[48] ENIAC was planned by Dr J. W. Mauchly and J. P. Eckert at the Moore School at the University of Pennsylvania, for the Ballistics Research Laboratory at the US Army's Aberdeen proving ground. It contained 18 000 valves and 1500 relays, weighed over 30 tons and was built by a team of 200 people between 1943 and 1945. Setting up a problem on ENIAC involved changing several hundred pluggable connections and an equivalent number of switches, and it was designed initially solely for ballistics calculations. Once set up, the machine had an impressive turn of speed with its 200 microsecond addition time and 3 millisecond multiplication time. During 1948 and 1949 ENIAC was modified by the introduction of a new system of coding and the addition of a mercury delay line to supplement its very small internal storage of only twenty numbers. ENIAC was then applied to several other problems including nuclear physics, and thus spanned the boundary between machines with and without an internal stored program.

Fig. 16.1 The shape of things to come. A dramatic view of the huge ENIAC electronic calculator, built towards the end of the second world war for ballistics calculations for the US Army. In the foreground an operator sets one of several hundred function switches prior to a computation run.

A second reason for the pause in American development was the mobility of personnel. At the Moore School, Eckert and Mauchly left the EDVAC project in the spring of 1947 to start their own company (eventually named the Eckert-Mauchly Computer Corporation). John von Neumann and Dr Herman Goldstine had already left EDVAC to return to the Institute for Advanced Study (IAS), Princeton. The EDVAC team was subsequently led by T. K. Sharpless, who himself then left to go into business.

It is possible that a third reason might have centred round the choice of practical storage devices. Talk of the RCA Selectron electrostatic storage system figured prominently in the early American reports, and its random access properties made it much to be preferred to the better-known mercury delay line. To quote Eckert from Lecture 45 of the

Moore School course (26 August 1946):

> Several forms of fast internal memory have been suggested and the one which shows promise at the present time is the electrostatic storage tube. The one on which most work is being done (by RCA) at the present time is called the Selectron. When it has been perfected it will contain most of the features which are desirable in this type of a memory.

The Selectron required an extremely complicated electrode structure and the device experienced such prolonged technical difficulties that many of the computer design teams hoping to use Selectrons decided to change their plans. Another shock came with the realisation that meanwhile a small British team had invented a successful random-access electrostatic storage system (the Williams tube), which was the subject of a number of patent applications in the hands of the National Research Development Corporation (NRDC). This was all the more galling because the Americans, including Eckert, had been working on electrostatic storage for some time. Several forms of American electrostatic storage such as the RCA Barrier Grid Tube were eventually developed, but none of the early ones had the regenerative simplicity of the Williams tube. Inevitably the NRDC found itself fighting in a complex patent interference action to defend British inventions and, on the brighter side, Williams tubes were used under licence in the IBM 701 and in several other American computers. Of course the balance was redressed eventually by the outstanding American invention of the magnetic core store. The principle of this was discovered in 1949 by Jay Forrester at the Massachusetts Institute of Technology (MIT). Core stores were first used in earnest in the American real-time high-speed computer project called WHIRLWIND in August 1953. (Whirlwind initially used electrostatic memory.)

The years 1950 to 1952 saw a veritable flood of American stored-program computers burst into life, some of which are described in more detail in Appendix 3. These include BINAC, CADAC, EDVAC, ERA 1101, MANIAC, ORDVAC, RAYDAC, SEAC, the IAS computer and the Eckert-Mauchly UNIVAC 1. This last machine, delivered in about

Fig. 16.2 UNIVAC I – the *tour de force* of the Presper Eckert/John Mauchly partnership in computer design. This machine was first installed in June 1951 and was remarkable for its advanced facilities for data handling. To the right of the photograph can be seen some of the magnetic tape decks, of which there could be up to ten.

June 1951, (and therefore later than the Ferranti Mark I in England), was the first to enter the American commercial scene. UNIVAC used mercury delay lines for its main storage and, amongst other interesting features, it had a very advanced magnetic tape system. There was by then a rapidly increasing sales potential in the USA with little competition for the UNIVAC in sight. In America, IBM had the resources to take up the challenge quite quickly (the IBM 701 was first delivered in 1953). In Britain it fell largely to the NRDC to try to promote awareness.

17
THE NRDC

One of the first jobs of the National Research Development Corporation after its foundation in July 1949 was to take over the Manchester computer patents, held temporarily by the Ministry of Supply. The NRDC was thus made acutely aware of the American sales prospects from an early date and took strenuous steps to encourage an alert and competitive British computer industry. NRDC set up the Advisory Panel on Electronic Computers, which, however, held but one meeting (on 14 December 1949). This was attended by all the major electronics and punched-card machine companies of the time. To quote reference 49:

> The outcome of the Advisory Panel meeting was that both the electronic manufacturers and the punched-card machine manufacturers respectively represented that they were individually in positions to tackle the problems of an electronic computer development project as well as, for example, the International Business Machines Corporation in the United States. It was pointed out to the punched-card machine manufacturers that, in the opinion of the Corporation, they had inadequate electronic staff and resources. It was apparent also that the manufacturers were not willing that the Corporation should take the initiative in launching a development project but agreed that the Corporation could usefully coordinate activities.

Despairing of a coordinated national effort, the NRDC turned to individual companies where, after a certain amount of coaxing, it had more success. As described earlier, Elliott Brothers were given support for the 401 development in 1950; Ferranti was supported for the Mark I Star in 1951 and for the PEGASUS in 1954; EMI Ltd received support for office computers in 1955, and installed a transistorised computer at the Austin Longbridge works in 1958. This was the prototype of the EMIDEC 1100 computer, first delivered in about 1959. This in turn was followed by the EMI 2400, designed at the request of the NRDC. The English Electric Company and BTM Company had meanwhile been progressing independently with NPL and Birkbeck College respectively. In parallel to this, the NRDC became aware that Britain was seriously lagging behind in the development of magnetic tape systems. Contracts were accordingly placed with several firms from June 1953 onwards.

Throughout the period the NRDC also pursued an energetic policy of patenting the fruits of computer research, spending over £100 000 on this activity up to 1956.[49] By that time it was administering 733 computer

Fig. 17.1 The spread of British know-how: left to right F. C. Williams, H. J. Crawley (NRDC) and J. C. McPherson (IBM) admiring the neat implementation of Williams electrostatic storage tubes for the IBM 701 computer in 1953.

patents in the UK and overseas, which had resulted from 201 inventions. The sources of these inventions were mainly Manchester University (81), NPL (43), The Ministry of Supply (36), and Ferranti Ltd (16). As an example of the resulting income, the total receipts from the sales and licensing of certain Manchester patents to IBM had amounted to £125 712 by the end of 1956. The University itself had received back about £27 000 by that date, under a special revenue-sharing agreement on Manchester patents. This money went towards further computer research.

As part of its policy of furthering development, the NRDC also retained a number of computer experts itself. Amongst these was Christopher Strachey. He did the bulk of the programming for the St Lawrence Seaway Project on a Ferranti Mark I computer at Toronto University (1952/3) and assisted with the 401 development when the 401 computer left Elliott Brothers (1953/4). From this grew PEGASUS, as described in Chapter 14.

Finally, in 1955 and 1956 the NRDC took an active part in the birth of what later became the British Computer Society. The 'computer professional' thereby came out into the open and the computer industry became a reality.

In conclusion, it must be said that the post-war economic climate in Britain was hardly conducive to investment in such a speculative area as stored-program computers. No one at that time could have predicted their subsequent importance. Indeed, there were many experts who declared computers to be of only marginal scientific interest. For example, in September 1951 Professor Douglas Hartree, in conversation with 'the' Ferranti computer salesman, could not understand the company's intention to market computers: 'We have a computer here in Cambridge; there is one in Manchester and one at the NPL. I suppose there ought to be one in Scotland, but that's about all'.[42] In contrast, such people as T. R. Thompson of the J. Lyons catering company were exceptional visionaries.

The decade 1945–55 is nevertheless remarkable for the high output of technological innovation and successful implementation from a comparatively small number of individuals. The British computer industry therefore had, and still continues to have, a good flow of ideas on which to feed. An analysis of the National Computer Index[50] indicates that by 1960 there were about 220 commercially produced computers installed in the United Kingdom. 90 per cent of these had been built by the nine British companies who at that time were offering general-purpose digital computers for sale. A decade later the number of installations had risen to about 6500 – of which only about half were designed and manufactured in the United Kingdom. There is a moral here somewhere!

APPENDIX 1

THE PLAIN MAN'S GUIDE TO COMPUTER TERMINOLOGY

A General concepts

A modern computer has four main sections which are conceptually linked as shown in figure A1.1 The *input* unit could be any device which converts letters, numbers and symbols into their electronic representation inside the computer. The simplest example is an electric *keyboard* similar to that used on a typewriter. Alternatively, information can be prepared on punched paper tape or punched cards, which are then fed into the computer via a *paper tape reader* or a *card reader*.

A1.1

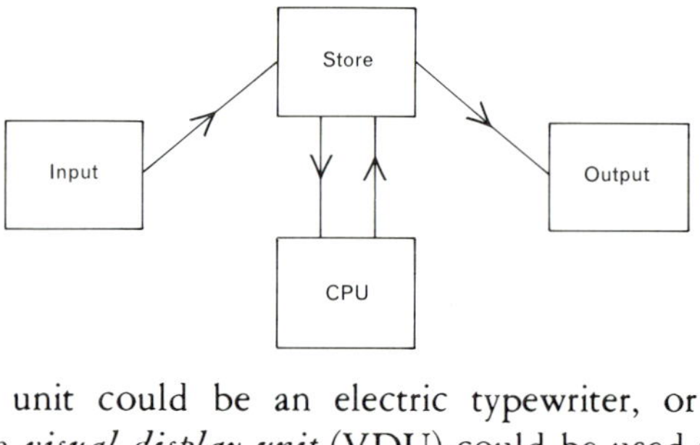

The *output* unit could be an electric typewriter, or a *lineprinter*. Alternatively, a *visual display unit* (VDU) could be used where a direct presentation of results is required.

The *CPU* or *central processing unit* is where the computation is actually carried out. It contains two sections: an *arithmetic and logic unit* (ALU) and a *control* unit. During execution of a program the CPU fetches instructions one by one from the store and obeys them. In principle an *addition* instruction, for example, could require the CPU to fetch from the store the quantities to be added, i.e. the *operands*, then perform the addition, and finally return the answer to a designated location. The time taken to perform a single addition instruction was usually about a *millisecond* (10^{-3} seconds) on the early computers. Later developments have achieved times of *microseconds* (i.e. 10^{-6} seconds) or even *nanoseconds* (10^{-9} seconds).

The *store*, or *memory*, is used to hold two types of information: firstly, the *program* or list of instructions prepared by the user; secondly, the *data* or facts which are required during the solution of a particular problem. The computer has no innate (in-built) intelligence, and in general it is necessary to feed in a fresh program and fresh data each time a new problem is to be solved.

The storage unit is crucial. Indeed, it is, historically speaking, the main distinguishing feature of the modern computer, when compared with other aids to calculation. In order to store and process information accurately and at high speed, the computer's store and CPU deal with *all* types of information in numerical, i.e. digital, form. Therefore, any information which, in the everyday sense, is initially non-numerical has to be converted to a digital representation before computation takes place.

A1.2

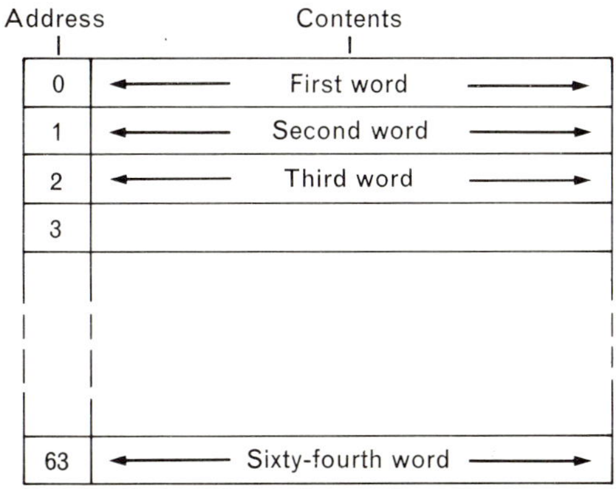

Bearing the above in mind, a computer's store may be imagined as many equal-sized pigeon-holes or *locations*, each one storing information as a string of digits. (Remember that everything is represented internally by digits.) Each store location is designated by a unique *address*, and the row of digits at each location is termed a *word*. Thus a simple picture of a 64-location store is shown in figure A1.2. Each word, which may represent either an *instruction* or *data*, is a string of digits. For reasons of engineering economy, computers use a *binary* (i.e. base two) system of numbers instead of our normal decimal (base ten) system. Examples are given below. Thus a word is a string of *binary digits* – abbreviated to *bits*. Several factors govern the choice of *word length* for a particular computer. An examination of a range of practical computer systems reveals that 16, 24, 32, 36 or 64 bits are popular sizes. A practical store may contain many thousand words of information.

Information may be transmitted to and from the store *serially*, i.e. one bit at a time, or in *parallel*, i.e. all the *n* bits of a word simultaneously down *n* wires. Similarly the arithmetic unit within a computer's CPU may process bits serially or in parallel. Serial mode is generally cheaper but slower. Most early computers were entirely serial; most modern ones are entirely parallel.

A1.3

Binary					Corresponding Decimal Value	
16	8	4	2	Units	Tens	Units
				0		0
				1		1
			1	0		2
			1	1		3
				:		:
	1	1	1	0	1	4
				:		:
1	1	0	0	1	2	5

The binary system of numbers, alluded to above, has the advantage of requiring just two symbols, – written as '0' or '1'. Binary arithmetic implies 'columns', headed units, twos, fours, eights, etc., in place of the more usual units, tens, hundreds, etc., of the decimal system. The principle of binary numbers and their everyday decimal equivalents is illustrated in figure A1.3. (Mercifully, a knowledge of the binary system is not normally needed when using a modern computer, since conversion from decimal to binary is carried out automatically!) Since a digital computer stores and processes *only* digits, any other non-numeric quantities such as letters of the alphabet have to be assigned a simple numeric code before being input to the computer. Once again, conversion of alphabetic symbols into coded binary, and vice versa, is carried out automatically.

B Variations in CPU design

Besides some obvious technological variations, different computers often employ different basic design strategies within their CPUs. This can be illustrated by describing two possible schemes or formats for representing computational instructions.

Scheme (a): *three-address instruction format*

In this scheme a computer word representing an instruction could be divided as illustrated in figure A1.4, showing a 32-bit word divided into four groups of eight digits. The *function* digits specify the operation to be performed; the two groups of *source* digits specify the store addresses of two operands, and the *destination* digits specify where the answer is to be placed in store. Thus an instruction may specify the equivalent of say:

ADD, 24, 73, 66

which is interpreted as: 'add together the numbers found in store locations 24 and 73, placing the result in location 66'.

A1.4

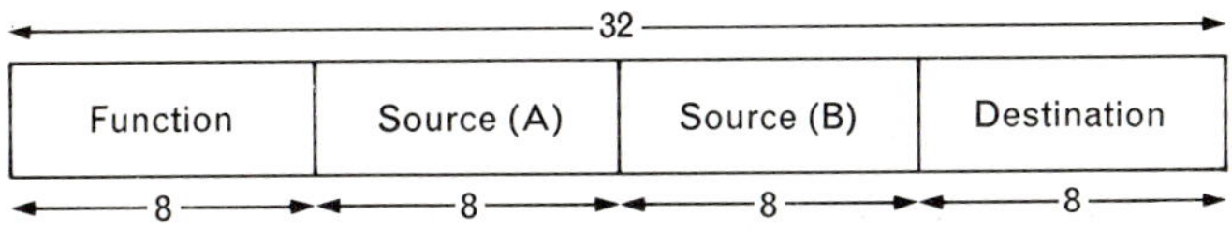

Scheme (b): *one-address instruction format*

An alternative layout for an instruction is shown in figure A1.5. (The partition 8/24 is for illustration and has no fundamental significance.) Each instruction now only carries one explicit operand address, though a second location is actually implied. This 'implicit' location is a one-word temporary storage place within the CPU called the *accumulator register*. (As a matter of terminology, a 'register' is generally used to denote any [temporary] storage location.) Using the one-address format, the previous addition of two operands to produce a result in store location 66 would now have to be programmed in three separate instructions, representing the equivalent of:

(*a*) load accumulator with the contents of address 24
(*b*) add into the accumulator the operand at address 73
(*c*) store the contents of the accumulator to address 66.

A1.5

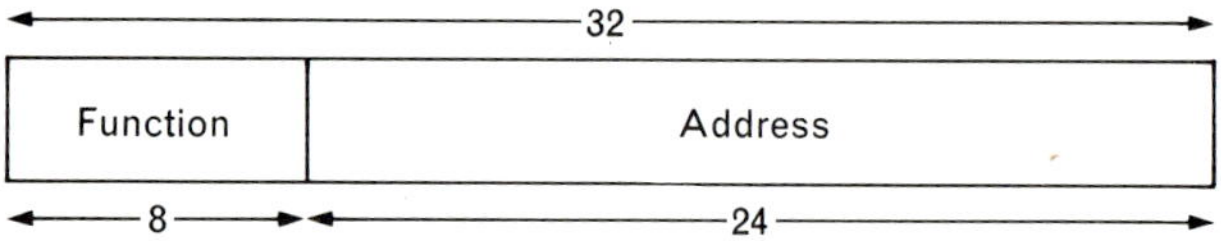

It is common nowadays to have computers whose instruction formats allow specification of one operand address, though two- or three-address formats are also seen. For those early computers which had essentially serial storage units, it was sometimes convenient to make each instruction also specify the address of the *next instruction* to be obeyed. (This 'next' instruction was not always the one placed consecutively in store.) In order to distinguish clearly the operand and next-instruction parts of a particular instruction format, a notation was often employed whereby the extra address used for specifying the location of the next instruction was denoted by '+1'. Thus a (2+1)-address instruction format provided specification for two operand addresses plus one next-instruction address; similarly for a (3+1)-address format, etc.

C CPU registers and structure

As has been said, a *register* is the general name given to any storage location (apart from the main store itself). Besides an accumulator, the CPUs of all modern computers contain another very important register known variously as the *program counter* or *control register*. This register holds the address of the next instruction to be obeyed. The contents of this register is normally incremented during execution of an instruction, so that the CPU fetches the next sequential instruction in a program. However the programmer may alter the program counter during computation, thus causing a *branch* or *jump* in the sequence of events. The programmer may thereby cause repetitive computation, or *conditional* computation.

Most computers also have one or more *index registers* (or *modifier* registers, or *B-lines*) within the CPU. These are used for a variety of reasons, but mainly for the formation of 'indexed' or 'subscripted' operand addresses. This greatly eases the task of handling structured data such as tables and matrices inside a computer. In a simple example of indexing, the final address of an operand could be formed by the automatic addition of a 'base' address as contained in the instruction plus an 'offset' or index as contained in a special index register. If there exist facilities for incrementing and testing the contents of this index register, it is easy to write a group of instructions which, when repetitively obeyed, causes the same computation to be performed on each item in turn in a list of items whose base address (i.e. the start of the list) is known. The invention of index registers in 1948 opened the way for other forms of offset address-generation in computer design.

Given that all CPUs contain a number of registers, the design philosophies for interconnecting and controlling registers fall into two (somewhat overlapping) categories: *hard-wired* control or *microprogram* control. For the former scheme, information about which register is connected where, and when, is inherent in the way the registers and their controlling circuits are wired together; under this scheme, any modifications to the CPU design involve basic changes at the wiring level. In contrast, a microprogrammed computer has a special (normally unalterable) *control store* which is microprogrammed with the information about register control. The CPU's structure is made to appear as a reasonably flexible arrangement of general registers and highways, the control of which may be expressed in terms of microprogram steps. A programming methodology may then be used when designing the sequence of operations needed during the execution of each instruction. Microprogrammed computers are in general easier to design and diagnose, but are slower than hard-wired computers.

D Computer software

The topics introduced so far have been related to computer *hardware*, and certainly the construction and operation of electronic circuits was the main concern of all early computer designers. Today a computer's *software* (i.e. the set of organisational programs provided by the manufacturers) is often more important to the user than the hardware.

Generally speaking a computer's software consists of three categories of program: *compilers, operating systems*, and *applications packages*. A compiler is a kind of translator which accepts a user's program written in a problem-orientated *high-level language* and translates it into the corresponding set of basic machine instructions, ready for execution. (*Fortran* is the name of one popular high-level language.) An operating system is the collection of housekeeping procedures which schedules the workload through a computing system, allocating resources and keeping accounts as necessary. Applications packages are programs, or collections of *subroutines*, for performing particular commonly occurring jobs. It is usual for a large computer installation to have a library of subroutines.

The programmer of an early computer had to have an intimate knowledge of its internal structure. Thankfully, such knowledge is not normally needed by the user of a modern machine. The connection between old and new programming techniques is further illustrated in Chapter 15.

Fig. A1.6 A cabinet of the Cambridge University EDSAC II computer (1957), showing the microprogram control store. The important concept of microprogramming was first expounded by Professor Maurice Wilkes of Cambridge University in 1951.

Appendix 2

TECHNICAL SPECIFICATION OF FIFTEEN EARLY BRITISH COMPUTERS

Table 3 gives the technical specification of six pioneering British research computers, arranged in chronological order according to their date of completion. (In terms of original design ideas, the Pilot ACE and MOSAIC should probably come first.) Since all of these computers were modified as time went on, the date corresponding to the given specification is important. This and other details are given below.

Manchester Mark I: The entry in Table 3 describes the machine which was operational in April 1949. Its 500-valve predecessor, the June 1948 computer, was a very small 32-bit word, 32-line store prototype with no index registers and only seven operations in its repertoire of instructions.[9] In the April 1949 version[45] the word-length was changed to 40 bits and a 30 millisecond revolution-time drum was added. Extra programming facilities included two index registers (B-lines) and a double-length (80 bit) accumulator. Note that the relatively long digit-period, a function of the Williams tube storage scheme, was offset from the user's view by the random access properties of this store and the provision of index registers. A contemporary performance evaluation rated the Ferranti Mark I at about the same usable speed as the NPL Pilot ACE, even though the latter had a digit period shorter by ten times.[9] Note that input/output transfers and drum transfers were initiated by operator switches, until program-control was provided in the October 1949 version of the Mark I. The Ferranti Mark I (1951) had eight index registers, addition and multiplication times of 1.2 milliseconds and 2.16 milliseconds respectively, a repertoire of 50 operations (see reference 57), and storage capacities of 256 words (CRT) and 4096 words (drum). In other respects it was the same as the October 1949 University Mark I.

Cambridge EDSAC: Table 3 describes the May 1949 version of EDSAC.[19] Certain engineering improvements were carried out in 1951 and some new control transfer orders were added in 1952 and 1953. Improvements of one form or another were continued over the next three years or so. A special feature of EDSAC was the wired-in 'bootstrap' or set of initial orders, which greatly eased the problem of cold-starting.

The descendant of EDSAC at Cambridge University was the EDSAC II computer,[51] first operational in 1957. This was a parallel machine with a 40-bit word and hardware floating-point facilities. It had a 1024-word ferrite core main store, a 1024-word core read-only memory, and magnetic tapes. Its fixed-point and floating-point addition times were respectively 20 and 75 microseconds, the corresponding multiplication times being 200 and 275 microseconds. EDSAC II used thermionic valves.

NPL Pilot ACE: Table 3 describes the state of the machine in December 1951. The 'number of operations' in the instruction repertoire is not easy to describe exactly, owing to the distributed internal structure of the computer as reflected in the instruction format. The figure of fifteen operations is given for purposes of comparison with more conventional machines having one-address instruction formats. The addition time could be anything from 64 microseconds to 1024 microseconds, depending on the position in the delay-line store of the next instruction. An average figure has been given.

The Pilot ACE had a complicated (2+1)-address instruction format which included provision for specifying one of 32 'sources', one of 32 'destinations' and the source of the next instruction. Instructions also specified the *duration* of a transfer, so that prolonging a transfer over several timing cycles could give the effect of shifting or multiplying operands by small integers. The instruction format permitted 'optimum coding', whereby the programmer could minimise the inconvenience of hold-ups due to the essentially sequential nature of delay line stores. He did this by suitable use of the 'next instruction source' field. During the initial design of ACE Turing was said to have been 'obsessed with speed of operation'.[52] To Turing, 'optimum coding' was just 'coding'. It was said that optimum coding could achieve a speed improvement of up to four times compared with conventional delay-line computers. In the long term the development of random-access stores rendered optimum coding

Fig. A2.1 Research on a shoe-string: the servo-synchronised magnetic drum store of the Manchester University Mark I computer in 1949. Because of its squat shape, the nickel-plated drum was referred to locally as the 'magnetic wheel'.

somewhat of a blind alley. A programming example is given in Chapter 15.

When compared with Alan Turing's original design for ACE,[12, 13] the actual Pilot ACE had a neater scheme for conditional branching. However, Turing had originally included an instruction address register and a return-link stacking/unstacking command for subroutines, and had been somewhat closer to a microprogrammed CPU design. It was not until 1958 that Dr A. M. Uttley, speaking at the inauguration of the full ACE, was able to say 'today Turing's dream has come true'.

NICHOLAS: Table 3 gives details of the computer working at the Elliott Brothers' Borehamwood establishment in December 1952. It was built in the Theory Laboratory by a team under Norman Hill and S. E. Hersom, initially for use by the laboratory for a series of trajectory calculations for a military project. NICHOLAS was based on the sub-miniature pentode and crystal diode circuits of the 153 project, the processor containing about 250 valves and 1500 germanium diodes.

Many of the logic boards were wired up at home in the evenings by members of the Theory Laboratory team.[31] The main store was made of eight nickel delay lines, each holding 128 words. An easy-to-use symbolic assembly language was developed for NICHOLAS. For example, although there was no hardware multiply instruction as such, a programmer could call upon a multiplication macro by simply writing the symbol 'M'. (The resulting instructions made use of a special conditional shift-and-add order, and were quite fast.) The computer gave valuable service until 1958.

TREAC: Details are given in Table 3 of the machine as described in March 1953 in reference 53. At that stage the magnetic drum had not been added but plans were well advanced. The drum had a moveable 24-head assembly which, when fixed, gave one track's worth of storage (2048 digits) per head at an average access time of 20 milliseconds. If greater capacity was required, the whole head assembly could be made to oscillate, producing a spiral effective track of 32 turns for each head and a total storage capacity of 64 000 words at a mean access time of 1.25 seconds. Since the whole computer was *parallel*, a figure for the digit-period is not quoted. In addition to the 2000 thermionic valves, TREAC contained 1000 germanium semiconductor diodes.

MOSAIC: This machine is as described in 1953.[54] The addition time could vary between 70 microseconds and 1.28 milliseconds and an average time is quoted. The paper tape input was three inches wide and came from a specially designed portable data recorder used in aircraft tracking experiments. In addition to the 6000 thermionic valves, MOSAIC contained 2000 germanium diodes.

Table 4 gives the overall characteristics of nine commercially available computers, as they existed in 1958. The information, which is in some cases only approximate, is mainly taken from reference 35. The IBM 704 is, of course, an American design, but is included for comparison. It is the fastest and most expensive computer described in Table 4.

Table 3 *Six early British research computers*

	Manchester Mark I	*Cambridge EDSAC*	*NPL Pilot Ace*	*Elliotts' NICHOLAS*	*TRE's TREAC*	*MOS's MOSAIC*
Word length, bits	40	36	32	32	24	40
Instruction length	20	18	32	16	24	40
Instruction format	1-address	1-address	(2 +1)-address	1-address	1-address	(3 +1)-address
Instruction set	26 ops	18 ops	(15)[a]	47 ops	14 ops	?
Store size, fast	128	512	352	1024	512	1024
Store type, fast	CRT	delay	delay	nickel	CRT	delay
Store size, backing	1024	0	0	0	2048[a]	0
Store type, backing	Drum	–	–	–	Drum	–
Add time (average)	1.8 ms	1.4 ms	0.54 ms[a]	12.5 ms	40 μs	0.67 ms[a]
Multiply time (average)	10 ms	5.4 ms	2 ms	a	–	6 ms
Input medium	PTR	PTR	Cards	PTR	PTR	Cards or PTR[a]
Ouput medium	Tprin	Tprin	Cards	Tprin	Tprin	Cards or Tprin
Digit period	8.5 μs[a]	2 μs	1 μs	3 μs	NA[a]	2 μs
Main valve type	EF 50	EF 54	ECC 81	VX8030/8046	?	CV 138
Approximate number of valves	1300	3000	800	250	2000	6000
Approximate number of Ge diodes	None	None	None	1500	1000	2000

Notation and conventions

a see text for further explanation.

Instruction set: the total number of useful operations in the repertoire is given.

Storage sizes: these are given in terms of full words.

Storage types: 'CRT' signifies Williams tube storage; 'delay' signifies mercury delay lines; 'nickel' signifies nickel magnetostrictive delay lines.

Speeds: 1 ms is 1 millisecond; 1 μs is 1 microsecond.

Input medium: 'PTR' signifies 5-track paper tape reader–(3-inch-wide tape for MOSAIC); 'Cards' signifies Hollerith punched cards.

Output medium: 'Tprin' signifies a teleprinter, which provided hard copy as well as 5-track paper tape.

Table 4 *Nine commercial computers available in 1958*

	Ferranti Mark I Star	*English Electric DEUCE*	*Elliott 402*	*BTM 1201*	*Lyons LEO II*	*IBM 704*	*Ferranti Pegasus*	*Metropolitan–Vickers MV 950*	*Ferranti Mercury*
Technology	Valve	Valve	Valve	Valve	Valve	Valve	Valve	Transistor	Valve
Word length	40	32	32	40	39	36	39	32	40
Store size, fast	416	400	17	0	1038	up to 32 763	56	(8)	1024
Store type, fast	CRT	Delay	Nickel	–	Delay	Core	Nickel	(Drum)	Core
Store size, backing	16 384	8192	5000	1024 or 4080	up to 65 536	8192 per drum	5120	4096	Up to 32 768
Store type, backing	Drum	Drum	Drum	Drum	4 drums	Drum	Drum	Drum	4 drums
FXPT add time	1.2 ms	32 μs minimum	204 μs minimum	1.25 ms	340 μs	24 μs	300 μs minimum	3 ms	60 μs
FLPT add time	–	–	–	–	–	72 μs	–	–	180 μs
FXPT multiplication time	2.16 ms	2.08 ms	3.366 ms	2–50 ms	0.6–3.5 ms	228 μs	2 ms	8 ms	210 μs
FLPT multiplication time	–	–	–	–	–	192 μs	–	–	300 μs
Date first delivered	1953	1955	1955	1956	1956	1956 in UK	1956	1956	1957
Approximate basic cost	90	50	27	37	90	500	50	?	100

Notation and conventions
These are mostly the same as in Table 3. In addition, 'core' signifies magnetic core stores; FXPT and FLPT signify fixed-point and floating-point arithmetic respectively. The 'minimum' addition times refer to computers with (1+1)-address instruction formats. On Mercury, index-register (B-line) arithmetic was faster than main accumulator arithmetic. The cost figures are in units of £1000, at 1958 prices.

APPENDIX 3

TECHNICAL SPECIFICATION OF NINE EARLY AMERICAN COMPUTERS

In general, most early American computers were faster and contained larger stores than contemporary British ones. A greater proportion of them were parallel and most included a division order in the instruction set. There were also some similarities, one of these being the choice of about 40 bits as a popular word-length on both sides of the Atlantic.

Many of the American computers followed one of two patterns: either they were based broadly on the final Institute for Advanced Study (IAS) design, characterised by electrostatic storage, parallel operation and one-address instruction format; or they followed the revised EDVAC design, characterised by mercury delay-line storage, serial operation and a (3+1)-address instruction format. There were of course exceptions, of which the UNIVAC was the most notable – (see Table 5 below). The SEAC was also unusual in that it initially had an EDVAC delay-line store and then acquired Williams tube electrostatic storage later. The electrostatic storage of all early American computers used the British Williams tube principle,[55] with the exception of the MIT WHIRLWIND computer which used a specially constructed double-gun tube designed by MIT.

The specimen stored-program computers listed in Table 5 were all considered 'large'. There were naturally many smaller ones. Notable amongst smaller machines was the Computer Research Corporation's CADAC,[56] designed by a group of engineers who had been working for the Northrop Aircraft Corporation on a digital differential analyser. CADAC was remarkable for packing a 1024-word drum store, 195 thermionic valves and 2500 germanium diodes into a box about one metre cube – which was provided with wheels! CADAC used a 42-bit word, a 3-address instruction format, and had an addition time of about 55 milliseconds. The first production machine was delivered to MIT in December 1951.

Fig. A3.1 The BINAC computer at the Northrop Corporation in America. Apart from References 44 and 59, few details of this historic machine seem to be available in the open literature, probably because of the classified nature of its application. Designed by the Eckert/Mauchly group, it is believed to have been delivered to Northrop in September 1949.

The nine large computers featured in Table 5 are arranged in approximate chronological order. Information about their characteristics mainly comes from references 4 and 8 and from appropriate papers presented at the conference cited in reference 55. Further background details are now given. Note that all the machines in Table 5 had hardware division instructions.

SEAC: 'Standards' Eastern Automatic Computer' was built in Washington DC by the American National Bureau of Standards, an organisation which in many ways serves the same purpose as the National Physical Laboratory in England. SEAC is distinguished from the somewhat later SWAC (Standards' Western Automatic Computer) built in Los Angeles. The entry in Table 5 refers to the first version of SEAC, working in May 1950 and probably the first American stored-program computer actually to become fully operational. The SEAC research team continued improvements during the next three years, adding parallel-access electrostatic storage and magnetic tape and magnetic wire.

BINAC: 'BINary Automatic Computer'. This machine was built by the company founded by J. P. Eckert and John Mauchly of ENIAC fame, for the Northrop Aircraft Corporation. It is believed to have been intended for airborne use, as part of a missile-control system. An interesting feature of BINAC was the duplication of the computational circuits and the main store, with a constant check being kept that the results of the dual calculations remained identical. If a discrepancy arose, the machine stopped. Input/output for the computer was initially under manual control. BINAC worked under test conditions in August 1949, and was shipped to Northrop in California the following month.[59] It is believed never to have functioned successfully at Northrop.

ERA 1101: This computer was designed by the Electronic Research Associates Company, primarily for undisclosed military applications. It is believed that ERA built several stored-program computers for strategic applications during the period under review. The ERA 1101 was first operational in December 1950.

WHIRLWIND: The WHIRLWIND project at the Massachusetts Institute of Technology (MIT) had its origins in wartime defence contracts for real-time aircraft simulation and anti-aircraft fire control. The requirements for high-speed calculations of defined accuracy led to the use of digital techniques and, though the application was initially a specialised one, the WHIRLWIND machine described in Table 5 was a true general-purpose stored-program computer. The entry in the table refers to the first version, operational in November 1950. This used a special MIT storage tube. Magnetic tape units and a ferrite core store were added later.

UNIVAC: 'UNIVersal Automatic Computer'. This machine was the *tour de force* of the Eckert-Mauchly team and in size, if not in speed, was the most impressive early American computer. It was designed very much with business data-processing in mind, unlike most other contemporary machines which were scientific in flavour. UNIVAC had hardware character-handling facilities and made provision for bulk input/output to magnetic tape in parallel with main computing activity. There were up to ten magnetic tape decks, called 'uniservos', using metal-backed half-inch-wide tape at a maximum transfer rate of about 21.6 Kilo-characters/second, using 60-word (720-character) fixed blocks. UNIVAC was delivered some time between March and June 1951.

ORDVAC: 'ORdnance Discrete Variable Automatic Computer'. This machine was built at the University of Illinois for the US Army's Ballistic Research Laboratories. Its general design followed that of the IAS computer at Princeton. ORDVAC used Williams tube storage and was first operational at the University of Illinois in November 1951. The later ILLIAC was very similar in design to ORDVAC.

EDVAC: 'Electronic Discrete Variable Automatic Computer'. The original 1945 proposal for this computer (see reference 10) was modified by the Moore School group in the University of Pennsylvania as time went by, in the light of available storage technology. Thus although the 1945 proposal is commonly regarded as the starting point for the design of the EDSAC, EDVAC and the IAS computer, the final products of these three design teams differed significantly. The EDVAC itself took rather a long time to be completed at the Moore School, being overtaken by nearly two years by its close derivative, SEAC. The entry for EDVAC in Table 5 refers to the machine which came into operation in April 1952.

IAS: John von Neumann's group at the Institute for Advanced Study, Princeton University, was the source of much inspiration to American computer design groups; one of the IAS-derived computers (at the Rand Corporation) was even named JOHNNIAC in recognition of this. Other computers based closely on the IAS machine were AVIDAC, ILLIAC, MANIAC, ORACLE and ORDVAC. The IAS computer featured in Table 5 was operational some time between January and June 1952; a drum store was added somewhat later.

RAYDAC: 'RAYtheon Digital Automatic Computer' was built by the Raytheon Manufacturing Company and installed at the Naval Air Missile Test Center, California. RAYDAC was unusual in having four digits in each word assigned as a check-sum, this being re-calculated and checked after each transfer. The arithmetic unit also produced a 5-bit check number for operands and result, and used these to carry out verification of each computation similar to the decimal system of 'casting out nines'. An unusual feature of the RAYDAC instruction set was hardware assistance for double-length and floating-point working. The entry for RAYDAC in Table 5 describes the machine which was operational in July 1952.

Table 5 *Nine early American computers*

	SEAC	*BINAC*	*ERA 1101*	*Whirlwind*	*UNIVAC*	*ORDVAC*	*EDVAC*	*IAS*	*RAYDAC*
Serial/parallel	Serial	Serial	Parallel	Parallel	Serial	Parallel	Serial	Parallel	Parallel
Word length (bits)	45	31	24	16	84	40	44	40	36[b]
Instruction length	45	14	24	16	42	20	44	20	72
Instruction format	(3 +1)-address	1-address	1-address	1-address	1-address	1-address	(3 +1)-address	1-address	(3 +1)-address
Main store size	512	512	16384	256	1000	1024	1024	4096	1024
Main store type	Delay	Delay	Drum	CRT	Delay	CRT	Delay	CRT	Delay
Backing store type	–	Mag. tape	–	–	Mag. tape	–	Mag. wire	(Drum)	Mag. tape
Add time (average)	864 μs	800 μs	96 μs[a]	49 μs	525 μs	72 μs	846 μs	62 μs	707 μs
Multiply time (average)	2.98 ms	1.2 ms	352 μs[a]	61 μs	2.15 ms	732 μs	2.9 ms	713 μs	868 μs
Basic clock frequency	1 MHz	1 MHz	400 KHz	1 MHz	2.25 MHz	Asynch.	997 kHz	Asynch.	4 MHz
Approximate number of valves	747	700	2700	5000	5400	2178	3600	2300	5200
Approximate number of Ge diodes	10500	?	2385	11000	18000	None	1000	None	17000

Notation and conventions

Serial/parallel: refers to the design of the arithmetic unit.

Main store type: 'delay' signifies mercury delay lines; 'CRT' signifies Williams tubes, except in the case of WHIRLWIND.

Clock frequency: ORDVAC and IAS were asynchronous computers, having no basic clock.

Ge diodes: these are germanium semiconductor diodes ('crystal diodes').

a The arithmetic times quoted for the ERA 1101 are *minimum* times; since the drum period was 10 milliseconds, maximum times were very long.

b The RAYDAC 36-bit word included four 'transfer-count' check digits.

BIBLIOGRAPHY AND REFERENCES

The serious enquirer into early British computers will get a very good idea of the spirit of the age by reading the proceedings of the first four major computer conferences. These were held respectively at the Royal Society in London in March 1948, the University of Cambridge in June 1949, the University of Manchester in July 1951, and at the National Physical Laboratory in March 1953. The proceedings of the first event were published in 1948 as A discussion on computing machines, *Proceedings of the Royal Society,* **195A**. Publication details of the other three are given below in references 19, 20 and 23. If a league table were to be constructed of the best-known British computer pioneers, it could probably be based on the number of times a particular individual appeared as author of a contribution to these four famous conferences. A fifth computer conference, held at the Institution of Electrical Engineers in London in April 1956 (see reference 26), describes amongst other things the commercial manifestation of many of the original research projects. It might thus be taken as signalling the end of the 'vintage' era of British computer design.

For more general reading, an excellent account of most early computer projects is given in B. V. (later Lord) Bowden's book *Faster Than Thought* (see reference 16 below). Lord Bowden is often referred to as the first computer salesman. A retrospective view by several eminent pioneers is provided by the collection of papers appearing in the July, August and October 1975 issues of the journal *Radio and Electronic Engineer* (see reference 18 below). From the historian's view these retrospective papers are sometimes disappointing, being a little short on hard facts and, in one or two places, a trifle inaccurate. Perhaps their greatest value is that they enable the modern reader to assess the personal philosophy and priorities of each pioneer. For the technical details there is no substitute for going back to the original reports, where these still exist.

Computer history started to become fashionable as a serious area of research in about 1975. In recent years a number of analytical articles have appeared in the journals (for example references 2, 6, 43 and 57). Such articles can be very valuable for interpreting long-obsolete terminology and for placing events in their scientific (if not social) context. For the latter, the London Science Museum has produced a series of archival tape-recorded interviews entitled 'Pioneers of Computing' which is very helpful. Finally, for those wishing to investigate the pre-1945 history of calculating machines, many original papers for the period 1834–1949 are reprinted in Professor Randell's book (see reference 10). This useful collection includes papers which are difficult to obtain, such as an account by Charles Babbage of his analytical engine.

REFERENCES

1 Morrison, P. & Morrison, E. (eds). *Charles Babbage and his Calculating Engines.* Dover, 1961.

2 Wilkes, M. V. Babbage as a computer pioneer. In *Proceedings of the Babbage Memorial Meeting, 1971* (British Computer Society, 1972). Reprinted in *Historia Mathematica,* **4**, 415–40, 1977.

3 Crank, J. *The Differential Analyser.* Longmans, 1947.

4 Goldstine, H. H. *The Computer from Pascal to von Neumann.* Princeton University Press, 1972.

5 Johnson, B. *The Secret War.* BBC Publications, 1978.

6 Randell, B. The Colossus. University of Newcastle Computer Science Technical Report 90, 1976. Reprinted in condensed form in *New Scientist,* **73**, 346–8, 1977.

7 Kilburn, T. A storage system for use with binary digital computing machines. Dept of Electrotechnics, University of Manchester, 1 December 1947. Later published (but without the section on a hypothetical computer) in Williams, F. C. & Kilburn, T., *Proceedings of the IEE,* **96**, Part 2, 183–202, 1949.

8 A Survey of Automatic Digital Computers, US Dept of Commerce, Office of Technical Services, Report PB111293, 1953.

9 Lavington, S.H. *A History of Manchester Computers.* National Computing Centre Publications, 1975.

10 von Neumann, J. Report on the EDVAC; mimeographed June 1945. Later incorporated in Burks, A. W., Goldstine, H. H. & von Neumann, J. *Preliminary Discussion of the Logical Design of an Electronic Computing Instrument,* Institute for Advanced Study, Princeton, June 1946. Reprinted in Randell, B. *The Origins of Digital Computers,* Springer-Verlag, 1973.

11 Turing, A. M. On computable numbers, with an application to the Entscheidungsproblem. *Proceedings of the London Mathematical Society,* Series 2, **42**, 230–65, 1936. Corrigenda in **43**, 544–6.

12 Turing, A. M. Proposals for the development in the Mathematics Division of an Automatic Computing Engine (ACE). Report E882, Executive Committee, NPL, February 1946. Reprinted April 1972 as NPL Report Com. Sci 57.

13 Carpenter, B. E. & Doran, R. W. The other Turing machine. *Computer Journal,* **20** No. 3, 269–79, August 1977.

14 National Physical Laboratory, Executive Committee minutes for Tuesday 21 January 1947.

15 Smith, R. A. Notes on a visit to the NPL to discuss the Automatic Computing Engine with the Director. TRE internal memo D.4070, 25 November 1946.

16 Bowden, B. V. *Faster Than Thought: A Symposium on Digital Computing Machines.* Pitman, 1953. (Contains useful material on most early British computers.)

17 Elliott, W. S. The present position of automatic computing machine development in England. In *Proceedings of the 2nd Harvard Symposium on Large-Scale Digital Calculating Machines, September 1949*, pp. 74–80. (Published 1951 as *Annals of the Computation Laboratory of Harvard University,* **26**.)

18 Wilkes, M. V. Early computer developments at Cambridge: the EDSAC. *The Radio and Electronic Engineer,* **45**, July 1975, 332–5.

19 Wilkes, M. V. The EDSAC. In *Report of a Conference on High-Speed Automatic Calculating Machines, Cambridge University, June 1949*, pp. 9–11, 1950. Reprinted by the Computer Laboratory, University of Cambridge, 1977.

20 Williams, F. C. & Kilburn, T. The University of Manchester computing machine. In *Inaugural Conference of the Manchester University Computer, July 1951*, pp. 5–11. (Also presented at the Joint AIEE/IRE Computer Conference, Philadelphia, December 1951.)

21 Knuth, D. E. & Pardo, L. T. The early development of programming languages. In Report STAN-CS-76-562 from Computer Science Dept, Stanford University, August 1976, pp. 41–7.

22 Kilburn, T., Edwards, D. B. G. & Thomas, G. E. The Manchester University Mark 2 digital computing machine. *Proceedings of the IEE,* **103** Part B, Supp. 1–3, 247–68, 1956.

23 Wilkinson, J. H. The Pilot ACE. In *Automatic Digital Computation* (Proceedings of a Symposium held at the NPL, March 1953), HMSO, 1954. (Reprinted in Bell & Newell, see reference 28 below.)

24 Woodger, M. The history and present use of digital computers at the National Physical Laboratory. *Process Control and Automation*, November 1958, 437–43.

25 Felker, J. H. Performance of TRADIC transistor digital computer. In *Proceedings of the Eastern Joint Computer Conference, Philadelphia, December 1954*, pp. 46–8.

26 Kilburn, T., Grimsdale, R. L. & Webb, D. C. A transistor digital computer with a magnetic-drum store. *Proceedings of the IEE,* **103B**, Supp. 1–3, 390–406, 1956.

27 Cooke-Yarborough, E. H., Barnes, R. C. M., Stephen, J. H. & Howells, G. A. A transistor digital computer. *Proceedings of the IEE,* **103B**, Supp. 1–3, 364–70, 1956.

28 Kilburn, T., Edwards, D. B. G., Lanigan, M. J. & Sumner, F. H. One-level storage system. *IRE Transactions on Electronic Computers,* **EC–11**, No. 2, 223–35, April 1962. (Reprinted in Bell & Newell, *Computer Structures*, McGraw-Hill, 1971.)

29 Coombs, A. W. M. An electronic digital computer, Parts 1–4. *Post Office EEJ,* **48**, 114, 137, & 212, July & October 1955 & January 1956; **49**, 18 & 126, April & July 1956. (Summary information also given in *Automatic Digital Computation*, pp. 38–42, see reference 54 below.)

30 The Telecommunications Research Establishment parallel electronic digital computer. Chapter 10 in *Faster Than Thought*. (See reference 16, above.)

31 Clarke, S. L. H. The Elliott 400 series and before. *The Radio and Electronic Engineer,* **45** No. 8, 415–21, August 1975.

32 Booth, A. D. Relay computers. In *Report of a Conference on High-Speed Automatic Calculating Machines, Cambridge University, June 1949*, pp. 17–21, 1950. Reprinted by the Computer Laboratory, University of Cambridge, 1977.

33 Lavington, S. H. An evaluation of the ARC project, Department of Computer Science, University of Manchester, Internal Report, May 1979.

34 Booth, A. D. The APE(X)C: a low-cost electronic calculator. In *Automatic Digital Computation* (Proceedings of a symposium held at the NPL, March 1953.) HMSO, 1954.

35 Vickers, T. Table comparing some digital computers available in Britain. *Process Control and Automation*, November 1958, 460.

36 Pinkerton, J. M. M. & Kaye, E. J. LEO – Lyons Electronic Office (part 1). *Electronic Engineering,* **29**, 284–91, 1954. Part 2 (entitled Operation and maintainance by Lanaerts, E. H.) **29**, 335–41. Part 3 (entitled A checking device for punched data tapes by Kaye, E. J. & Gibbs, G. R.) **29**, 386–92.

37 Haley, A. C. D. The inconspicuous computer. *The Radio and Electronic Engineer,* **45**, 409–10, August 1975.

38 Haley, A. C. D. DEUCE: a high-speed general-purpose computer. *Proceedings of the IEE,* **103B**, Supp. 1–3, 165–73, 1956.

39 Allmark, R. H. & Lucking, J. R. Design of an arithmetic unit incorporating a nesting store. *Proceedings of the IFIP Congress, Munich, 1962*, pp. 694–8. (Reprinted in Bell & Newell, see reference 28.)

40 Davis G. M. The English Electric KDF9 computer system. *Computer Bulletin*, December 1960, 119–120.

41 Elliott, W. S., Owen, C. E., Devonald, C. H. & Maudsley, B. G. The design philosophy of Pegasus, a quantity production computer. *Proceedings of the IEE,* **103B**, Supp. 1–3, 188–96, 1956. (Reprinted in Bell & Newell, see reference 28.)

42 Swann, B. B. An informal history of the Ferranti Computer Department, 1975. (Circulated privately.)

43 Campbell-Kelly, M. A series of papers on early British programming systems (EDSAC, Manchester Mark I, etc.) to appear in the *Annals of the History of Computing*, a quarterly journal published by the American Federation of Information Processing Societies.

44 Auerbach, A. A., Eckert, J. P., Shaw, R. F., Weiner, J. R. & Wilson, L. D. The Binac. *Proceedings of the IRE*, 12–29, January 1952.

45 Kilburn, T. The Manchester University digital computing machine In *Report of a Conference on High-Speed Automatic Calculating Machines. Cambridge University, June 1949*, pp. 119–21. Reprinted by the Computer Laboratory, Cambridge University, 1977.

46 Newman, M. H. A. Some routines involving large integers. In *Report of a Conference on High-Speed Automatic Calculating Machines, Cambridge University, June 1949*, pp. 69–70. Reprinted by the Computer Laboratory, Cambridge University, 1977.

47 Brooker, R. A. An attempt to simplify coding for the Manchester electronic computer. *British Journal of Applied Physics,* **6**, 307–11, September 1955.

48 Goldstine, H. H. & Goldstine, A. The Electronic Numerical Integrator and Computor (ENIAC), *MTAC* (now *Mathematics of Computation*) 2, No. 15, 97–110,

1946. (Reprinted in Randell, see reference 10 above.)

49 Crawley, H. J. The National Research Development Corporation Computer Project. NRDC Computer Sub-Committee Paper 132, February 1957.

50 *The National Computer Index*, maintained by the National Computing Centre. (Data for the early years compiled by Derek Pedder.)

51 Renwick, W. EDSAC II. *Proceedings of the IEE,* **103B**, Supp. 1–3, 204, 1956.

52 Wilkinson, J. H. The Pilot ACE at the National Physical Laboratory. *The Radio and Electronic Engineer,* **45**, 336–40, 1975.

53 Carter, R. H. A. The TRE high-speed digital computer. In *Automatic Digital Computation* (Proceedings of a Symposium held at NPL, March 1953), pp. 56–64. HMSO, 1954.

54 Coombs, A. W. M. MOSAIC – the Ministry of Supply Automatic Computer. In *Automatic Digital Computation*, (Proceedings of a symposium held at NPL, March 1953), pp. 38–42. HMSO, 1954.

55 Forrester, J. W. Digital computers: present and future trends. In *Review of Electronic Digital Computers*, a joint AIEE/IRE Computer Conference held at Philadelphia, December 1951, pp. 109–13. (This conference also contained papers on UNIVAC, ORDVAC, ERA 1101, WHIRLWIND and SEAC.)

56 Sprague, R. E. The CADAC. In Proceedings of a symposium on commercially available general purpose electronic digital computers of moderate price, Washington, May 1952, pages 13–17. (Sponsored by the Navy Mathematical Computing Advisory Panel.)

57 Lavington, S. H. The Manchester Mark I and ATLAS: a historical perspective. *Communications of the ACM,* **21**, 4–12, January, 1978.

58 Denison, S. J. M. Further DEUCE interpretative programs and some translating programs. In *Annual Review in Automatic Programming*, ed. R. Goodman, vol. 1, pp. 127–45, Pergamon, 1960. GEORGE was implemented by C. L. Hamblin; see references in his paper Translation to and from Polish notation, *Computer Journal,* **5**, No. 3, 210–13, 1962.

59 Stern, N. The BINAC: a case study in the history of technology. *Annals of the History of Computing,* **1**, No. 1, 9–20, July 1979.

60 Barton, R. S. Programming languages and computer design: a case history. In *High-Level Languages*, Infotech State of the Art Report No. 7, pp. 363–74. Infotech Information Ltd, 1972.

ACKNOWLEDGEMENTS

The active encouragement of the Science Museum, South Kensington, London is acknowledged in the preparation of this book. Thanks are due to Professor Brian Randell for his helpful comments on the manuscript.

The author is indebted to many people for their personal recollections. Particular mention should be made of B. Bellringer, S. L. H. Clarke, Dr A. W. M. Coombs, H. J. Crawley, L. Dilger, Professor W. S. Elliott, A. C. D. Haley, S. E. Hersom, Dr S. H. Hollingdale, A. St Johnston, Professor Tom Kilburn, D. Pedder, Dr J. M. M. Pinkerton, C. J. Somers, Professor A. M. Uttley, Professor Maurice Wilkes, Dr J. H. Wilkinson, and M. Woodger. Notwithstanding their willing assistance, the responsibility for the text rests finally with the author. It will also be appreciated that the recording of 'recent history' is not an exact science, especially where it impinges on personalities and national security!

PICTURE CREDITS

Bassano & Vandyk Studios, London: Figure 8.1

S. L. H. Clarke: Figures 11.1, 11.2, 11.3

Professor A. J. Colin: Figure 12.1

Dept of Computer Science, University of Manchester: Figures 2.4, 4.1, 4.2, 4.3, 4.4, 4.5, 4.7, 7.1, 7.2, 7.4, 9.1, 9.2, 9.3, 9.4, 14.1, 15.1, 15.3, 15.4, A2.1. (With thanks to Mike Koiston, photographer.)

Ferranti Limited: Figures 7.3, 7.5, 14.2, 15.2

R. Jordeson, University of Manchester: Figure 2.3

National Physical Laboratory (Crown Copyright): Figures 4.6, 12.2, 12.3, 13.2, cover photo.

National Research Development Corporation (H. J. Crawley): Figure 17.1

Dr J. M. M. Pinkerton: Figures 6.3, 13.1

Post Office Research Centre, Martlesham Heath: Figure 10.1

Public Record Office (Crown Copyright): Figures 3.1, 3.2

Science Museum, London (Crown Copyright): Figures 1.1, 2.1, 2.2, 5.2, 13.2, 16.1, A3.1, cover photo.

Sperry Univac Computer Systems: 16.2

United Kingdom Atomic Energy Authority, Harwell, (R. C. M. Barnes): Figure 12.4

Professor A. M. Uttley: Figure 10.2

Professor M. V. Wilkes, University of Cambridge: Figures 6.1, 6.2, A1.1

M. Woodger, National Physical Laboratory: Figure 8.2

INDEX

152 and 153 fire-control systems, 57–8, 116
(2+1)-address instruction format, definition of, 110; examples of, 92, 115, 118

accumulator register, definition of, 109–10
ACE (Automatic Computing Engine), original proposals for, 25–30, 43, 44–6, 116; final computer, 46, 116; pilot model, *see* Pilot ACE
acoustic delay lines, *see* mercury delay lines
address, basic concepts of, in computers, 107, 109–10
Admiralty, 32, 56, 58; Signals Establishment, 32
aircraft, computers applied to, 8, 53, 66–7, 122
All-Power Transformer Co., 53
Allmark, R. H., 76
Allway, G. G., 46
analogue computers, 56, 66
Analytical Engine, *see under* Babbage
anticipation pulse effect, 19
APE(R)C computer, 28, 62–4
ARC (Automatic Relay Computer), 62, 65
assembly language, 33, 90, 117
Atanasoff, J. V., 7
ATLAS computer, 16, 37, 51, 52, 81, 83–5
Atomic Energy Authority (UKAEA), Harwell, 24, 49, 65, 67
autocode of Glennie, 41; Mark I, 42, 95
AVIDAC computer, 123

B-lines, *see* index registers
Babbage's, Charles (1791–1871), 2, 126; Babbage Analytical Engine, 2
Barnes, R. C. M., 65
barrier grid storage tube, 18, 100
Barton, S. A., 31
Baudot code (teleprinter code), 9; *see also* teleprinter
Bell Telephone Laboratories, 14, 48, 49; calculators 98
Bennett, J., 31
Berry, C., 7
BINAC computer, 92, 94–7, 100, 121, 122, 125
binary number system, explanation of, 108
Birkbeck College, University of London 24, 62–4, 85, 103
bit, definition of, 9, 108
Blackett, P. M. S. (later Lord Blackett), 38, 39
Bletchley, *see* Government Code and Cipher School, Bletchley
Booth, A. D., 62–5, 85
Booth, S. J., 63
Borehamwood Laboratories, Elliott Brothers, 24, 57–9, 116
Bowden, B. V. (later Lord Bowden), 126
British Computer Society, 104
British Rayon Research Association, 62
British Rubber Producer's Research Association, 62
British Tabulating Machine Co. (BTM), 63, 64, 69, 73, 81, 83–5, 103; 1200 series computers, 64, 82, 85; 1201 computer, 64; 1300 series computers, 86
Britten, Miss Kathleen, 62
BTM, *see* British Tabulating Machine Co.
Brooker, R. A., 42, 51, 91
Brunsviga calculators, 4
Brunt Committee, 50
Burroughs Corporation, 76; B5000 computer, 76
Bush, Vannevar, 5

CADAC computer, 100, 120
Cadby Hall (J. Lyons Co., London), 24, 68, 71, 72
CADET computer, 49
calculators, electro-mechanical, 4, 6, 23, 98
Cambridge University, 2, 5, 23, 24, 26, 29–30, 31–4, 53, 60, 68–70, 104, 113, 115, 126; EDSAC computer, *see* EDSAC; Mathematical Laboratory, 29, 34
cards, *see* punched cards
cathode ray tube stores, 17–19, 26–7, 36, 114, 118, 119
CDC 6600 computer, 81
central processor, description of, 106, 109–11
chip, silicon (integrated circuit), 12, 15–17, 49
Chisholm, K., 75
Coales, J. F., 57
Cockroft, Sir John, 52
code deciphering, use of computers for, 8–11, 53
COLOSSUS, 8–12, 23, 53; Mark II, 11
compilers, 41; definition of, 112
Computer Research Corporation, 120
Comrie, L. J. (1893–1950), 6, 23
conditional branching, 11, 111, 116
Cooke-Yarborough, E. H., 65
Coombes, A. W. M., 23, 25, 26, 47, 53
core store, principle of, 21–2; computer use of, 100, 115, 119, 122
Coventry Gauge & Tool Co., 35, 71
CPU (central processing unit) description of 106, 109–111
Crawley, H. J., 103
crystal diode, definition of, 14; *see also* diode, semiconductor technology

Darwin, Sir Charles Galton (1887–1962) Director of NPL from 1938, 25–7, 29
Davies, Donald, 26
decatron valve, description of, 13; application of, 65–6
delay line stores, general principles of 17–18, 21, 36; applied to computers 17, 27, 71, 93–4, 115, 118–19; *see also* mercury, nickel
Delpht University, Holland, 66
Department of Atomic Energy, 27–8
Department of Scientific and Industrial Research (DSIR), 27, 46
DEUCE computer, 17, 19, 47, 66, 74–6, 79, 92, 119
differential analyser, 5–7; electronic, 66; meccano, 6; digital, 120
digital computer, principles of, 1, 106–12
Digital Equipment Corporation, 61
Dilger, L., 66
diode, semiconductor, description of, 14–15; use of, 53, 116–18, 125; thermionic, description of, 13–15
disc, *see* magnetic disc stores
drum, *see* magnetic drum stores

Eckert, J. Presper, 98, 99, 100–101, 121, 122
Eckert–Mauchly Computer Corporation, 99
EDSAC computer, 18, 28, 29, 31–5, 44, 53, 68–70, 76, 90, 94–7, 115, 118, 123; EDSAC II, 34, 113, 115
EDVAC computer, 32, 99, 100, 123, 125; report 24, 26, 46, 95, 98, 99, 120, 123
Edwards, D. B. G., 37, 40, 51
Electronic Research Associates (ERA), 122; *see also* ERA 1101 computer
electrostatic stores, general principles of, 17–19, 21; applied to computers, 17, 19, 36, 100, 120, 122; *see also* Williams tubes, Selectron
Elliott Automation (company), 77
Elliott Brothers (manufacturing company), 18, 56–61, 84, 102, 104, 116; 400 series computers, 57; 401 computer, 58, 60, 79, 102, 104; 402 computer, 60, 62, 82, 119; 405 computer, 60, 61; 800 series computers, 61, 85; 802 computer, 61; 803 computer, 52
Elliott, W. S., 58, 60, 79–80
EMI (Electrical and Musical Industries Ltd), 44, 84, 102; EMIDEC 1100 computer, 52, 103; 2400 computer, 103

English Electric Co., 47, 68, 74–7, 84, 103
English Electric Leo Computers Ltd., 74, 77
English Electric Leo Marconi Computers Ltd., 77
English Electric Computers (company), 77
'English Electric' computers, *see* DEUCE, KDF9 etc., System 4
ENIAC, 3, 23, 98–9
Enigma codes and machines, 8–11
ERA 1101 computer, 100, 122, 125

Farmer, P., 31
Farnborough, *see* Royal Aircraft Establishment
Ferranti Ltd., 38, 39, 41, 51, 78–86, 102, 104; ARGUS computer, 81, 82, 85; ATLAS computer, 16, 37, 51, 52, 81, 83–5; FP6000 computer, 81, 86; Mark I computer, 16, 19, 39–43, 61, 78, 81, 91, 93, 104, 114; Mark I Star computer, 39–40, 78, 102, 119; MERCURY computer, 43, 78, 119; ORION computer, 81, 83–6; Packaged Computer, 79; PEGASUS computer, 17, 57, 60, 78–82, 86, 102, 104, 119; PERSEUS computer, 78; POSEIDON computer, 83; SIRIUS computer, 52, 85
Ferranti, Sebastian de (1864–1930), 78
Ferranti–Packard Company (Canada), 86; FP6000 computer, 86
flip-flop (bistable circuit), description of, 16–17
floating-point arithmetic, 43, 115, 119, 124
Flowers, T. H. (Tommy), 11, 23, 25, 26, 47
Forrester, J. W., 21, 100
Fort Halstead, 40–41
Fortran, 89, 112
FPC1 (Ferranti Packaged Computer), 79

Geheimschreiber codes and machines, 8–11, 42
General Electric Co. Ltd. (GEC), 84, 86
GEORGE (General Order Generator), 76
Gill, S., 34
Glennie, Alec E., 40–41, 91
Gold, T., 32
Goldstine, H. H., 35, 70, 99
Government Code and Cipher School, Bletchley Park, Buckinghamshire, 8–11, 23–5, 42, 47, 53
GPO (General Post Office) *see* Post Office
Grimsdale, R. L., 48
guided weapons, computers applied to, 38, 53, 66, 122, 123

Haley, A. C. D., 74, 76
Halsbury, Lord, 50
hardware, general definition of, 86, 111
HARRIAC computer proposal, 86
Hartree, Douglas R. (1897–1958), 5–6, 23, 29, 34, 70, 104; wife of, 26
Harvard University calculators, 98
Harwell, *see* Atomic Energy Authority
de Havilland Propellers Ltd., 35
HEC (Hollerith Electronic Computer), 63, 64
Hersom, S. E., 116
high-level programming languages, 95, 112; *see also* autocode, Fortran
Hill, Norman, 116
holding-beam storage tube, 18
Hollerith, Herman, 4, 64
Hollerith punched-card equipment, 44, 63
Hollingdale, S. H., 65
Hopper, Grace, 26
Howarth, David, 51
hybrid computers, 66

IAS *see under* Princeton University
IBM Corporation, 84, 102, 103–4; 701 computer, 100, 101, 103; 704 computer, 117, 119; card-programmed electronic calculator, 98; selective-sequence electronic calculator, 98; STRETCH computer 51; System/360, 370 series 77, 82
ICCE computer, 65–6
ICL (International Computers Ltd), 77, 84–6; 1900 series computers, 80, 81, 82, 85–6; 2900 series computers, 86; System 4 computers, 77

ICT (International Computers & Tabulators), 77, 83–6; 1500 series computers, 85, 86; 1900 series, *see under* ICL
ILLIAC computer, 123
Illinois University, 123
Imperial College, University of London 24, 65–6
index registers, principle of, 111; in computers, 37, 38, 56, 81, 95–7, 114, 119
Institution of Electrical Engineers (London), 126
Iowa State College, 7

Jacquard loom, 2
JOHNNIAC computer, 123
Johnson, H. (of Ferranti Ltd), 86

KDF6 computer, 76; KDF7 computer, 75; KDF8 computer, 76; KDF9 computer, 76–7, KDN2 computer, 75; KDP10 computer, 76
Kelvin, Lord (William Thomson), 5
Kidsgrove, English Electric factory, 24, 74, 76
Kilburn, Tom, 26, 28, 36, 37, 39, 40, 41, 43, 48, 50–2, 86, 88

Lanigan, M. J., 52
LARC computer, 50, 51
Lenaerts, E. H., 70
LEO computer, 28, 33, 35, 68–73; LEO Computers Ltd., 35, 74, 84; LEO II, 35, 74, 119; LEO III, 74; LEO *Chronicle,* 65, 69–73
Lincoln Laboratory, MIT, 49
Lockspeiser, Sir Ben, 38, 39
London University, 62, 64, 65
Lonsdale, Keith, 41
Lyons, J., Co. Ltd., 33, 35, 68–74, 104

machine-code (programming), 89–94, 96, 112
McPherson, J. C., 103
MADM, *see* Manchester Mark I
magnetic disc stores, general principles, 21; in computers 57–8, 60
magnetic drum stores, general principles, 20–1; in computers, 37, 48, 50, 56, 62–6, 74, 80, 91, 97, 98, 114, 116, 117, 118–19, 120, 123, 125
magnetic film stores, general principles, 21; in computers 60
magnetic tape stores, general principles, 21; in computers, 35, 69, 74, 80, 101, 103, 122, 123, 125
magnetic wire store 122, 125
magnetostrictive delay lines, *see* nickel delay lines
Malvern, Worcestershire (site of TRE), 8, 24, 53–4
Manchester Mark I computer (MADM), 28, 37–40, 43, 51, 56, 90–92, 94–7, 114, 116, 118; prototype (1948) or 'baby machine', 36, 114
Manchester Mark II, *see* MEG
Manchester University, 3, 5, 23–5, 27–30, 36–42, 48–52, 78, 83, 86, 104, 126; experimental transistor computer 48–50; *see also* ATLAS, MU5, MUSE
MANIAC computer, 100, 123
Marconi Co., 77, 84
MARK I, *see* Ferranti Mark I or Manchester Mark I
Marks, Cecil, 26
Massachusetts Institute of Technology (MIT), 21, 49, 100, 120, 122; *see also* WHIRLWIND
Mauchly, John W., 98, 99, 100–101, 121, 122
MEG computer, 43, 48
memory, *see* store
MERCURY computer, *see* Ferranti MERCURY
mercury delay lines, 17–18, 25, 32, 44, 53, 74, 98, 99, 101, 120, 125; *see also* delay line store
Mersenne prime numbers, 37
Metropolitan–Vickers Co., 48, 51
Michaelson, S., 65
microcomputer, 90
microprogram, 34, 94, 111, 113, 116

microsecond, definition of, 13, 107
millisecond, definition of, 13, 107
minicomputer, 61
Ministry of Supply, 23, 35, 53, 54, 72, 102, 104
missiles, computers applied to, *see* guided weapons
MIT, *see* Massachusetts Institute of Technology
modular circuit techniques (packaging), 57, 58, 60–61, 79–80
Moore School of Electrical Engineering, University of Pennsylvania, 23, 25–6, 98–100, 123
morphogenesis, 30, 93
MOSAIC computer, 23, 28, 53–4, 114, 117, 118
MRS5 real-time fire control system, 57, 60
MU5 computer (Manchester University), 16, 86
MUC (Manchester University Computer), 37, 42
MUSE computer, 50, 51–2
MV950 computer 48, 51, 62, 119

National Bureau of Standards (America), 121; *see also* SEAC, SWAC
National Cash Register Co. (NCR), 84
National Computer Index, 105
National Computing Centre, 85
National Physical Laboratory (NPL), 23–30, 43–7, 53, 68, 74–5, 103, 104, 121, 126; Electronics Division, 44; Mathematics Division, 23, 26, 29, 44
National Research Development Corporation (NRDC), 40, 50, 58, 60, 79, 100, 101, 102–4
Naval contracts (British), 57–58
Nelson, Sir George (later Lord Nelson), 74
Neumann, John von (1903–1957), 23, 25, 62, 95, 99, 123
Newman, E. A. (Ted), 44, 46, 74
Newman, M. H. A. (Max), 11, 23, 24, 26, 29, 37
NEWT computer, 86
NICHOLAS computer, 28, 57–9, 116–17, 118
nickel delay lines (magnetostrictive), 17–18, 57–8, 80, 117, 118–19; *see also* delay line stores
Nieman, H. W., 5
Northrop Aircraft Corporation, 120, 121, 122
Norwegian Defence Research Establishment, 43
NPL, *see* National Physical Laboratory

one-address instruction format, explanation of, 109–10; example of, 89–90
Owen, C. E., 57
optimum programming, principles of, 94; use of, 79, 115–16
ORACLE computer, 123
ORDVAC computer, 100, 123, 125

paging 19, 52
paper tape, use of, 9, 34, 106; equipment 9, 33, 37, 53, 80, 81, 89, 118
parallel computers, definition, 108; examples, 54–5, 56, 62, 117, 120, 125
parity-bit, 80
patents 37, 103–4
PDP/8 computer, 61
PEGASUS computer, *see* Ferranti PEGASUS
Pennsylvania University, 23, 25–6, 29, 98, 123; *see also* Moore School
pentode valve, description of, 13, 14
Petherick, E. J., 65
Philco Corporation, 49
Physical Society Exhibition 58
Piggott, R., 31
Pilot ACE, 28, 29, 44–7, 74–5, 92–4, 114, 115–16, 118
Pinkerton, J. M. M., 35, 70, 72
Plessey Co., 66
van der Poel, W. L., 66
point-contact transistor, 15, 48, 49
Pollard, Brian, 41
Post Office Research Station (Dollis Hill, London), 8, 11, 25–7, 47, 53–4

Powers, James, 4
Powers–Samas Accounting Machines Ltd., 84
Princeton University, 23–5, 35, 62, 70, 98, 99; Institute for Advanced Study (IAS), 99, 120, 123; IAS computer, 100, 123, 125
Prinz, D. G., 38
processor (and CPU), description of, 106, 109–11
programming systems, 33–4, 87–97, 112, 117; *see also* assembler, autocode, high-level language, optimum programming, subroutine
punched cards, use of, 2, 94, 106; equipment 4–6, 44, 69, 73, 75, 81, 84–5, 102, 118

Queen's University, Belfast, 5

radar, computer techniques in, 8, 53, 58
Radio Corporation of America, *see* RCA
Rand Corporation, 123
random access, principle of, 36, 37
RASCAL computer, 66
RAYDAC computer, 100, 123–5
Raytheon Manufacturing Co., 123
RCA, 24, 76–7, 99–100; 301 computer, 86; 501 computer, 76; SPECTRA computer, 77
register, definition of, 109–11
relays, principle of operation, 13–14; applied to computers, 11, 63, 65–6, 98
Renwick, W., 31
Robinson, A. A., 40
Rothamstead Experimental Station, 60
Royal Aircraft Establishment, Farnborough, 5, 24, 65–6, 74
Royal Signals and Radar Establishment, 8
Royal Society, 45, 126

Science Museum, London, 46, 126
Scientific Computing Service Ltd., London, 6, 23
SEAC computer, 100, 120, 121–2, 123, 125
SEC (Simple Electronic Computer), 62
SECA computer, 66–7
SEDA computer, 66–7
selectron storage tube, 18, 23–4, 99–100
semiconductor technology, definitions, 13–16, 22; *see also* diode
sequence-controlled calculators, 7, 62, 65
serial computers, basic definition of, 108
Sharpless, T. K., 99
Smiths Aircraft Instruments Ltd., 66
Smith, R. A., 26–7
software, general definition of, 86, 111–12
St. Johnston, A. (of Elliott Brothers), 58
Standard Telephones & Cables Ltd., 66, 69, 71–3
Stantec ZEBRA, *see* ZEBRA
stack, concept of, 76
Stevens, G. J., 31
storage technology 16–22; terminology 107–8
store (or memory), basic concept, 106–108; *see also* core, delay line, electrostatic, magnetic disc, etc.
stored-program computer, definition of, 1; basic operation of, 106–10
Strachey, Christopher, 79, 81, 104
STRETCH computer, 50, 51
SWAC computer, 121
subroutine libraries, 34, 87, 95, 112
subroutine-calling mechanisms, 95, 116
Sydney University, Australia, 76
System 4 computer, 77

tape, *see* magnetic tape or paper tape
Telecommunications Research Establishment (TRE), 8, 24–9, 36, 37, 54–6; *see also* Malvern
teleprinter, computer applications of, 9, 11, 33, 37, 42, 66, 90, 118
thermionic valves, principle of operation, 13–14; applied to computers, 11, 16, 33, 37, 40, 44, 46, 52, 53, 57, 62, 63, 65, 78, 114–19
Thomas, G. E. (Tommy), 37, 40
Thompson, T. Raymond, 26, 35, 69, 70, 72, 104

Thomson, James (1822–1892), 5, 7
three-address instruction format, explanation of, 109
Tocher, K. D., 65
Tootill, G. C., 36, 39
Toronto University, 104
TRADIC computer, 48
TRANSAC S-1000 computer, 49
transistor computer, experimental, Manchester 48–50
transistors, description of, 14–16; application to computers, 48–52, 61, 84, 86, 103, 119
TRE, *see* Telecommunications Research Establishment
TREAC computer, 28, 54–56, 117, 118
Turing, Alan M., 10, 23, 25, 28–30, 42, 43, 44–6, 93, 115–16
two-address instruction format, definition of, 110
two-plus-one address format, definition of, 110; examples of, 92, 115, 118
TX-0 computer, 49

ultrasonic delay lines, *see* mercury delay lines
UNIVAC Corporation, 51, 84; UNIVAC computer 100–101, 120, 122–3, 125; LARC computer, 51
universal computer, definition of, 1
Uttley, A. M., 26–8, 54, 116

vacuum tubes, *see* thermionic valves
valves, *see* thermionic valves
virtual storage 19, 52

Wayne Kerr Laboratories Ltd., 35, 71
Wheeler, David J., 26, 31, 34
WHIRLWIND computer, 100, 120, 122, 125
Wilkes, Maurice V., 26, 29, 31–5, 44, 70, 113
Wilkinson, J. H., 26, 44, 46, 74, 93
Williams, Sir Frederic C., 26–8, 36, 37, 39, 40, 43, 91, 103
Williams tubes, 17–19, 21, 36, 37, 55–7, 100, 114, 118, 120, 123, 125
WITCH computer, 67
Wolverhampton College of Technology, 67
Womersley, J. R., 29
word (computer), definition of, 33, 107–8

Z3 computer, 7
ZEBRA computer, 66
zero-address computers, 76
Zuse, Konrad, 7, 26; KG (manufacturing company), 7